G. V. Black

The Formation of Poisons by Micro-Organisms

G. V. Black

The Formation of Poisons by Micro-Organisms

ISBN/EAN: 9783744715973

Printed in Europe, USA, Canada, Australia, Japan

Cover: Foto ©berggeist007 / pixelio.de

More available books at **www.hansebooks.com**

THE

FORMATION OF POISONS

BY

MICRO-ORGANISMS.

A BIOLOGICAL STUDY OF THE GERM THEORY OF DISEASE.

—

BY

G. V. BLACK, M.D., D.D.S.

PHILADELPHIA:

P. BLAKISTON, SON & CO.,

No. 1012 WALNUT STREET.

1884.

PREFACE.

The historical portion of this volume has been condensed from notes made from time to time in the study of the subject. With the view of making it as short as possible, I have given only those experiments and observations that seemed important to a proper understanding of the subject, avoiding all but the most necessary details; aiming at the same time to preserve all that will be of real advantage to the general student. Many familiar names will be missed, for the reason that many have written, and some of them very well indeed, who have developed no distinctive fact or thought that is of service in the farther progress of the subject. When we come to analyze any such subject, most men are surprised to find how few have been instrumental in the development of the real basic facts on which our knowledge of it rests. Therefore, if we can properly estimate the import of the matter presented, the history of the development of any such subject may be briefly written. Whether or not I have determined wisely, the intelligent reader must judge.

The second part was written because I had something to say that I thought ought to be said at the present time. The men who have been most instrumental in the development of this subject have given us little else than the experimental facts. These will satisfy the minds of but very few. Most of us appreciate any subject more if we understand why these things are so. It is this *why* that I have sought to supply. This I leave for intelligent discussion to sift, and separate the good grains from the chaff.

G. V. BLACK, M. D., D. D. S.

JACKSONVILLE, June 2d, 1884.

TABLE OF CONTENTS.

SECOND PART.

FOURTH LECTURE.

FIFTH LECTURE.

SIXTH LECTURE.

SEVENTH LECTURE.

APPENDIX.

PART FIRST.

ACKNOWLEDGMENT.

***

This volume contains a series of lectures, delivered before the students of the Chicago College of Dental Surgery and a number of practitioners who were ticketed especially for this course. The Lectures now appear in their present form through the co-operation and kindly assistance of the Officers and Faculty of that Institution, to all of whom I gratefully acknowledge myself indebted for many favors.

THE AUTHOR.

A STUDY

OF THE

GERM THEORY OF DISEASE.

HISTORICAL.

INTRODUCTION.

In presenting an historical sketch of the Germ Theory of Disease, it will be my effort to review briefly the ideas, controversies and experiments, that have gradually led up to our present knowledge concerning it. At present I know of no single treatise on this subject from which those who have not followed its now extensive literature can gain a clear view of it, obtain a just conception of its magnitude, of the labor that has been bestowed upon it, or of its immense importance. The history of this subject in its completeness, would be a history of the efforts of men to stay the ravages of epidemic and contagious diseases; a history of the efforts of men to understand the causes of plagues and pestilences; a history of the efforts of men to see farther, and still farther, into the secret causes which produce sickness and diminish the longevity of the race. I cannot, however, give more than a brief synopsis of the principal thought, experiment and discussion, bearing most directly upon, and finally leading to the adoption of present theories. The literature of

2   11

the subject is now quite extensive, and is, for the most part, scattered through works upon disease in general, more especially works on surgery. Very few books were written especially on this branch of medicine until quite recently; and these generally dealt with some particular phase of the subject, or with the very recent experimentation, giving little or no account of preceding inquiries. Therefore it is very difficult for one who takes up the subject now to gain a comprehensive view of it as a whole. It will be my object to supply such a view in a short and concise recital of the thought, experiment and discussion, that has been most effective in leading to our present knowledge, and most essential to an understanding of the work now being done in this field.

FIRST TRACES.

For many centuries these ideas consisted in vague conjectures, arrived at from the study of contagion—an undefined something that could pass from the sick to the well, and cause disease. We find that Ulysses (Homer's Odyssy, Book XXII), used sulphurous acid to destroy the odor and toxic products of decomposition.

Early in the history of Greece we find that men had learned that certain localities were unhealthy. Their notion seemed to be that something obnoxious to health was being generated at these places; and we find them acting with judgment in regard to the location of their hospitals and important buildings. The physicians of antiquity observed that, in epidemic diseases the then existing theories were insufficient. They recognized that there was something extraordinary to deal with. They spoke of a "constitutio pestilens" and of a "genus epidemicus," but as to the nature of this constitution of disease they had few clear ideas.

Diodorus found an explanation of the cause of the Athenian

Plague in the circumstance that "a great multitude of people from all quarters streamed into the city, and being cramped for room, breathed corrupted air." Decomposing filth, social squalor, bad weather, etc., were regarded as causes of disease, by both physicians and laymen.

INCENTIVES TO THE STUDY.

When we turn our attention to the contagious diseases, to the great plagues of past times, and watch the progress of thought in regard to them, the absence of knowledge of the means of staying their progress, the utter helplessness of the people in times of pestilence, seems simply terrible. Without a knowledge of the diseases of a country, we find it difficult to understand aright its history or its civilization. These great diseases have often destroyed the army of the conqueror, or given the death blow to an advancing civilization, and have left a strange and enduring impress upon the intellectual life of great communities. It is generally known how, in the Fourteenth Century, the most deadly of all the pestilences that are recorded in past history—the Black Death—changed the direction of intellectual and social activity throughout the chief part of the civilized world, and showed its impress on the developments of succeeding centuries. We can gain but a faint idea to-day, through what we have seen around us, of the devastation that may be caused by epidemic diseases running without check ; or of the significance they have had in the progress of civilization.

The terrible results of epidemic and contagious diseases have furnished the strongest possible incentives to the study of the underlying causes. In all ages of the world the effort to understand them has been unceasing.

FIRST DEFINITE ANNOUNCEMENT.

The first *definite* announcement of a belief that disease is caused by organic germs, that I have found, was in the time of the Roman Empire.

De-re-Rustica, Varro and Columella, refer the origin of malarious fevers to the entrance of low organisms into the body. But they seem not to have given the special observations upon which their conclusions were founded. All along down the centuries, the idea was frequently expressed that the plague of the day was caused by minute organisms.

This doctrine, however, obtained wide recognition when some sort of basis for such theories was furnished by the microscopic demonstration of very minute living organisms, invisible to the naked eye. Especially after the discovery of the spermatozoa by Leuwenhœck in 1677. These were then, and for years afterwards, supposed to be real animals. It having now been demonstrated, apparently, that real animals were living in the bodies of men, the doctrine that diseases were caused by minute organisms spread far and wide. Among the best known advocates of this theory were Kirchen, Lancici, Valisneri, Raumum, and Linne. But even those who best understood the theory never reached anything more than rough conceptions; while many lost themselves in wild exaggerations. The animals causing disease were described as flying about in the air, something like swarms of insects, with crooked bills and sharp claws; and one writer proposed to destroy them during epidemics by the blowing of horns and the firing of cannon. We can readily conceive that such fantasies would bring down ridicule upon the whole theory; and in time these wild notions were dissipated and the germ theory slept.

PHLOGISTON.

But amid these fantasies there was a deep vein of earnest observation and thought. Physicians saw a pestilence strike a community as a spark strikes among shavings, and kindling, attack person after person in society, as fire would leap from house to house in a city. Sydenham says : " Ita ignis ignem generat et maligno infectus morbo socium inficit."

Stahl supposes a principle of inflammability or the matter of fire in composition with other bodies, an hypothetical element supposed to be pure fire, fixed in combustible bodies, in order to distinguish it from fire in action or in a state of liberty. This was called " Phlogiston." A similar supposition obtained in regard to epidemic and contagious diseases, viz., that there was some substance or force in man himself, which, when once put in motion, acted similarly to fire and spread with deadly effect. To this day, remedies directed against *inflammation* or *inflammatory* conditions are called antiphlogistics. This theory of the origin and propagation of epidemic diseases seems to have been much discussed.

ZYMOSIS.

Another hypothesis was based on the known action of ferments, which is maintained to this day, the present germ theory being considered within the limits of this hypothesis. It was seen that yeast, when added to certain compounds containing sugar, caused certain phenomena. That a very small quantity of the yeast was necessary for a beginning, and that this would be reproduced continuously. It could be carried from vat to vat of grape juice for any number of times, and the more fermentation accomplished the more the ferment increased. It was found, also, that this kind of fermentation would begin, but less promptly, without the addition of yeast, and that in this case also the yeast was formed. Now they reasoned that the action of epidemic and contagious diseases was the same. From the sick person it seemed that a substance or force was conveyed to the well. Thus, one sick person may infect a dozen, and each of these a dozen more, etc., so that the disease is rapidly diffused through a community. It will be seen at once that the similarity between the phenomena of these diseases and the phenomena of fermentation is very close. These two hypotheses gave rise to much discussion. They seem to have stood side by side until the one merged into the other, for the more they were discussed the more similar they seemed to become. Finally, the first is lost sight of and the Zymotic theory of epidemics and contagions, as developed by Willis in 1659, has continued to be the generally accepted theory.

JENNER.

When Jenner discovered the relation of the vaccine disease to smallpox, and began the use of the vaccine virus as a protective against that malady in 1798, there came a much greater confidence in the fermentation, or as it is called, the zymotic theory of contagion. Here the physician had a body that he could carry about and use at will, as he could yeast. He could introduce this disease-producing yeast, or virus, into the skin of a healthy person, and after a certain time, a disease would manifest itself; just as he could introduce yeast into a solution of sugar, and after a certain time, find the phenomena of fermentation. A small amount of either was sufficient; in each case a certain period of rest was observed before the characteristic manifestations declared themselves; a stage of incubation. In each case the peculiar product first used was largely increased. In case of the vaccine, a dozen could be vaccinated from the product of the first vaccination, just as from the product of the fermentation of a single butt of must enough yeast could be produced to quickly start fermentation in a dozen more; and so on, *ad infinitum,* in both cases. The phenomena of the one following precisely the role of the phenomena of the other in every respect, only one was the fermentation of sugar, and the other was the production of the disease, Vaccina.

Neither did the likeness stop here; but, as we shall see, it went much further. It was then well known that a certain amount of heat destroyed the power of yeast forever. Trial showed that precisely the same thing happened with the vaccine virus. Furthermore, it was well known that when a solution of sugar has been under the process of fermentation until that process has spontaneously ceased, a further fermentation could not be had by adding more yeast; although unfermented sugar may remain in the solution. The results when completed are completed once for all; and for a new trial a new mixture must be had. Or in other words, the result of fermentation prevents fermentation of the same character again taking place. These phenomena are again repeated by the vaccine virus. A person once vaccinated is rendered insusceptible to further vaccination; at least for some years. All these facts taken together furnished the strongest possible proof of the identity of the two processes, at least in mode of operation; however wide the difference in the results. And the well known fact, that in the great majority of the then known contagious diseases, one attack rendered the subject of it incapable of again taking the same malady, served to extend the theory to all diseases of this class.

NINETEENTH CENTURY.

This is the condition in which we find the theory of zymotic diseases at the beginning of the present century. To say that this theory was universally accepted and satisfactory, would be untrue. Very many denied the relevance of the facts brought forward, and sought other explanations; but without gaining much favor. Earnest men were examining the act of fermentation; and especially seeking the causes of putrefaction. Why should a dead body pass so quickly into a state of putrefaction? Enough was known to convince many thoughtful men that putrefaction was not caused by anything inherent in the flesh itself; for, under some circumstances, decomposition did not take place. They had learned to prevent it, at least for a long time, in meats that were preserved for food. Many other circumstances came up which furnished food for thought, and which shook the faith in the idea that fermentation of vegetable substances and putrefaction of animal tissues were truly spontaneous and inherent in the nature of such bodies.

EARLY DISCOVERIES.

In the first decade of the sixteenth century, Von Helmont had shown that the gas arising from fermentation was different from common air ; was carbon dioxide. And the other product, alcohol, had been known from the earliest historic times, but it was not isolated until the fifteenth century, though its *distillation* was regularly practiced in the eighth. It had been shown by Gay-Lussac, that grape juice does not ferment in vacuo. This was discovered in a now classical series of experiments, in which this experimenter caused clean grapes to ascend through the mercury of a large barometer into the Toricellian vacuum, and then crushed them by means of a mercurial column. This juice remained unchanged, but the addition of small quantities of air set up fermentation. Stahl, in 1731, arrived at the conclusion that fermentation and putrefaction were similar or identical processes. This was also the opinion of Justus Liebig, one hundred years later.

In 1680 the Dutch professor Leuwenhœck began the examination of yeast under the microscope, and found it to consist of minute ovoid or globular particles. But the imperfection of the instruments of his time prevented the discovery, which seemed so near, of the nature of these particles. Experiments of various kinds were instituted by many persons, a few of which may be mentioned : A bladder was filled with a solution of sugar and suspended in another solution of sugar, to which yeast had been added. While fermentation went regularly on in the fluid in the containing vessel, the fluid in the bladder did not partake in the fermentation. A vessel was divided into two compartments by a partition of filtering paper, a solution of sugar placed in each and yeast added to one. Fermentation went on promptly in the compartment to which the yeast was added, but the other remained free from fermentation. Many other experiments are also recorded, all tending to show that fermentation is something different from ordinary chemical action. But the men of the time seemed unable to understand the significance of their experiments, which, to us, prove conclusively that the substance causing fermentation is particulate and not soluble, otherwise it would pass through the filters and produce its results.

In 1787 Fabroni affirmed that " the matter which decomposes sugar is a vegeto-animal substance; it resides in particular utricles in grapes, as well as in corn. When grapes are crushed this glutinous matter is mixed with the sugar. Directly the two substances come in contact, effervescence and fermentation commence."

Astier, in 1813, asserted that " the matter of ferment, recognized by Fabroni as an animal substance, was alive and derived its nourishment from the sugar, whence resulted the rupture of the equilibrium between the elements of this body. By this theory," said he, " it is easily explained that all the

2*

causes which kill animals or hinder their development must be opposed to fermentation " (Schutzenberger, page 38). This seems to be the first appearance of the modern germ theory in a definite form, and the first announcement of the true theory of the antiseptics. Astier, however, regarded the living substance as composed of animalcules. A number of observers within the next few years seem to have arrived at the same conclusion.

SCHWAN.

The next real advance was made about the year 1838, when Schwan and Latour, each independently, took up the microscopic inquiry, with improved instruments, and discovered that the granules of yeast were membranous bags, which exhibited all the morphological characters of vegetable cells, and under proper conditions increased and multiplied, in the biological sense. The conclusion was quickly reached by them, that it was the life and growth of the plant which caused the chemical changes of fermentation ; and that the products, carbon dioxide and alcohol, were excrementitious products of the plants.

At this time Schwan was working on spontaneous generation, which he very distinctly negatived. He also, together with Schultz and Helmholtz, was the first to establish the fact that *putrefaction* would not occur without the presence of certain minute organisms, which might be destroyed by boiling ; and then, that air deprived of these organisms by being heated, or by filtering through sulphuric acid, might be admitted and yet no decomposition would occur. This destroyed the hypothesis previously held, that *oxygen* was the active agent in decomposition ; yet it was not generally accepted. It is to this discovery that we owe our ability to preserve fruits, meats, etc., in cans hermetically sealed, which has conferred such blessings upon mankind.

EFFECTS OF SCHWAN'S DISCOVERIES.

These brilliant discoveries very soon attracted general attention to the subject from the scientific world, and as identical theories, as we have shown, had long been held as to the modes of propagation and action of contagious diseases, and of fermentation and putrefaction, the theory of a contagium vivum again came to the front, and was urged by some of the strongest minds of the time. The new theory, however, was destined to fight its way inch by inch. Every new fact that was put forth had to pass the most rigid criticism that the opponents of the theory could bring to bear upon it; and for a time it seemed as if it would be crushed out of existence notwithstanding its apparent demonstration.

LIEBIG'S OPPOSITION.

The strongest opponent of the vital theory, as it seems to me, was the then comparatively young Professor of Chemistry at Giessen, Germany, Justus Liebig. Liebig's view of fermentation is practically the same as that enunciated by Willis in 1659, and maintained by others in the succeeding years; but it is in Professor Liebig's writings that the view reached its highest stage of development and can be best studied. It merits the closest scrutiny, for it has been the principal opponent with which the germ theory has had to contend. If we follow the contests between the chemists on the one side and the vitalists on the other, up to the present time, and analyze the arguments adduced by the chemists, we will find that the very few new facts which they have brought forward have all been based on the plan of Liebig's arguments. There has been nothing essentially new, nor has his argument been strengthened. In the years 1840–42, in response to an invitation of the British Association for the Advancement of Science, Professor Liebig wrote a series of papers upon animal,

vegetable and agricultural chemistry and the chemical trans-
formations, including the action of the different ferments,
yeast, putrefaction, contagion, miasm, etc., in which he antagonized the vital theory of each of these with all the vigor of his
wonderful intellect, and for the time crushed the rising confidence in the new theory.

The arguments brought forward by Liebig may be summed
up as follows : "Fermentation is a result of the catalytic action
of a decomposing body in contact with compounds of feeble
molecular affinity, which is brought about in accordance with
the following law of dynamics : A molecule set in motion by
any power can impart its own motion to another molecule
with which it may be in contact." Yeast, he argued, is a product of the decomposition of gluten, and is necessarily a decomposing body, and when added to must or wort, sets up in
these bodies a motion of their molecules similar to its own, by
which their saccharine elements are converted into the simpler
and more stable compounds, carbonic acid and alcohol.

But the action upon the gluten is to convert it into a body
identical with the original yeast. And thus the yeast seemed
to grow ; but it is not a growth in the biological sense. This,
he contended, is demonstrated by adding yeast to a pure solution of sugar, in which case, although the fermentation proceeds promptly, the quantity of the yeast not only did not
increase, but actually diminished, being expended in the act
of decomposition.

Putrefaction, he says, is just the same process, but with a
different molecular motion. These molecular motions might
reproduce themselves or not, as the substance on which they
acted contained the substance from which they were produced or not. Miasms and contagions were considered as
being of the same character.

His idea of a chemical force more powerful than the vital

force appears, also, in his explanation of the manner in which poisons and medicines act. The manner in which inorganic poisons (medicines) gain admission to the tissues is owing, in most cases, to the formation of a chemical compound by the union of the poison with the constituents of the organ upon which it acts; it is owing to a chemical affinity more powerful than the vitality of the organ. Each of these substances, in its transit, produces a peculiar disturbance in the organism; in other words, they exercise a medicinal action upon it, but they, themselves, suffer no decomposition. It is only when the solutions are diluted to a certain degree with water that they are absorbed by animal tissues. In respect to this physical property of the animal tissues, alcohol resembles the inorganic salts.

Miasms and contagions were thus explained: "A miasm is a form of molecular force developed during the decomposition of vegetable bodies under certain peculiar circumstances, which has the power of setting up similar decompositions in the human blood; but not finding in the blood or tissues the proper substance for its own reproduction, like yeast in sugar, it is not reproduced. Therefore, such diseases are not contagious."

"Contagious poisons find in the human blood and tissues substances from which they reproduce themselves, as yeast in must or wort is reproduced from the gluten; consequently, these diseases are conveyable from the sick to the well; in some instances by contact only, as in syphilis, in which case the physical form of the substance is that of a solid. In other instances through the air, in which case the physical form is that of a gas." He says, "several kinds of contagions are propagated through the air; so that, according to the view already mentioned (of a contagium vivum), we must ascribe life to a gas, that is, to an aeriform body."

He thus recognized a gaseous form of ferments.

We have already spoken of the fact that when fermentation spontaneously ceased, it could not again be set up; and its application to those contagious diseases which can, ordinarily, be had but once by the same individual. This theory was elaborated anew by Liebig, with telling effect against the vital theory. It was considered that in must or wort there is a certain substance that is decomposed by the peculiar molecular motion of vinous fermentation, namely, sugar; while the other portions of the compound are not necessarily affected. The blood is a very complex substance, containing many compounds held together by feeble molecular affinities, any one of which may be decomposed. In case of the contagious diseases, the decomposing body which causes the particular disease acts upon some single compound of the blood, decomposing it. If this particular component of the blood be very important to the vital processes, the disease will be very grave. If, on the other hand, this particular compound be not very important, the disease will be correspondingly light. Now in the progress of the disease this particular compound is all decomposed, as the sugar in grape juice, and if the patient has been able to withstand the shock, he returns to health with this one compound of the blood lacking, and it often happens that it is replaced very slowly, if at all; therefore, the patient is protected for a time or permanently.

·Liebig explained the susceptibility of children to certain diseases while adults were insusceptible, by assuming that the constituent of the blood upon which the molecular motion poison acted in the child was absent in the adult. He also explained the differences in susceptibility of different individuals in the same way.

Finally, Liebig says, "all the supposed proofs of the vitality of contagions are merely ideas and figurative representations,

fitted to render the phenomena more easy of apprehension by
our senses, without explaining them. These figurative ex-
pressions, with which we are so willingly and easily satisfied
in all sciences, are the foes of all inquiries into the mysteries
of nature; they are like the Fata Morgana, which show us
deceitful views of seas, fertile fields and luscious fruits, but
leave us languishing when we have most need of what they
promise. It is certain that the action of contagions is the
result of a peculiar influence dependent upon chemical forces,
and in no way connected with the vital principle."

Liebig claimed that this poisonous substance could not be
isolated, because it consisted of a peculiar form of molecular
motion. But in many cases the substances containing this
molecular motion poison could be had, which substance in
itself was not poisonous, as was shown by exposing it to heat,
thus rendering it innocuous by destroying this peculiar molecular power.

He also points out the fact that very many of these poisons
are destroyed in the human stomach in the act of digestion,
and that others are not. He explains these phenomena in
this way. All of these ferment poisons owe their power to
the peculiar molecular movement started and maintained by
the act of decomposition. Now, when they come in contact
with stable compounds they are unable to communicate their
own movements to these associated molecules, and are consequently inert, or unable to propagate decomposition. But,
if the intermixed substance is composed of molecules held
together feebly, the molecular movement is communicated to
them, and they pass into a state of decomposition.

Now the juices of the stomach are acid, and among the
noxious substances many are alkaline, but some are acid.
It therefore happens that when an alkaline organic poison or
ferment is introduced into the stomach, its molecular force is

antagonized by the acid juices it meets there and its molecular movement destroyed.

If, however, the noxious element be acid in its reaction, the juices of the stomach do not interfere with its molecular movement, but rather favor it. And instead of being destroyed, it is rendered even more virulent. This action of the stomach, he argues, is a very positive proof of the chemical nature of these poisons.

It is a noteworthy fact that in this essay Prof. Liebig practically ignores the microscopic demonstrations of the yeast plant. He evidently regarded it as, to a large extent, a fiction of the brain of over-zealous investigators, using what was in that day considered very high powers of the microscope. Not believing in the existence of the yeast plant, or the organisms of disease, he says, "The vital principle is only known to us through the peculiar forms of its instruments; that is, through the organs in which it resides. Hence, whatever kind of energy a substance may possess, it is amorphous and destitute of organs from which the impulse, motion or change proceeds; it does not live."

The basis which he gives for the proof of the existence of life stands to-day in its full power. And if he is at last beaten in the argument, it is because facts have been brought forward and proven by incontrovertible observation that will satisfy this basis. This basis is contained, practically, in one sentence, which I quote: "Our notion of life involves something more than mere reproduction, namely, the idea of an active power exercised by virtue of a definite form, and production and generation in a definite form."

DIFFICULTIES.

From that time to the present the controversy has been unceasing. Those who believed that the vital theory was true still maintained it, for they had seen definite forms reproduced in definite forms, by virtue of definite forms. They returned to their microscopes and followed again the growth of the yeast plant. They saw again the buds given off from the cells, and reproduce other cells, like in every respect to the mother cell; and these in turn reproduce other cells of the same likeness; satisfying the basis as given by Prof. Liebig.

But the microscopes of that day were poor, trained observers were scarce, the subject was new, the modes of demonstration were comparatively clumsy, and in many respects faulty. Under these difficulties progress was exceedingly slow, at best; and for a time no substantial advance was made. But gradually, as the number of the observers of it increased, the yeast plant, as the active principle of yeast, and upon which its action was dependent, was accepted by a large proportion of the learned men of the world. It is, however, a noteworthy fact that many of those who accepted this refused to accept the vital principle for the other fermentations; which would seem naturally to follow. This refusal seems to have found a basis in certain arguments brought forward by Prof. Liebig, especially such as the following: If oxamide be brought in contact with oxalic acid dissolved in water, the following changes take place: the oxamide is decomposed by the oxalic acid, provided the necessary conditions for their exercising an action on each other be present. Although the oxamide is not at all soluble, the elements of water unite with it, and ammonia is one product formed and oxalic acid the other, both in the proportions to form a neutral salt. Here the contact of the oxamide and the oxalic acid induces a transformation of the oxamide, which is decomposed into oxalic acid and ammonia. As much oxalic acid exists after the decomposition as was added, and still possesses its original power; and more oxamide will be decomposed in the same way, if added. This may be kept up continuously, and any amount of the oxamide decomposed, by a very minute portion of the oxalic acid.

Quite a number of similar chemical decompositions are known, and while they cannot be considered as bodies in a state of decomposition, as Liebig claimed the excitors of fermentation were, the processes had long been considered as very closely related. Now as these processes were admitted by all to be purely chemical, no process should be considered vital until it was definitely proven.

This required that each one of these be investigated on its own merits and proven independent of all others.

In this work the theory has, perhaps, suffered as much from its over-zealous followers, who hastily published untenable views, arrived at by hasty and ill-digested experiments, as from its opponents.

SCHROEDER.

Schroeder was, perhaps, the first to make a new and substantial point that has stood the test of adverse criticism. As we have stated, Schwan had demonstrated that decomposition would not occur in sterilized fluids upon the introduction of air that had been heated or passed through sulphuric acid. This, as a demonstration of the existence of life, was apparently demolished by Liebig, who claimed that the heat, or the acid either, was sufficient to destroy the molecular motion upon which, as he claimed, the power of ferments depended. Now Schroeder, in 1854, admitted air filtered through cotton batting, with the idea that this would catch any solid particles in the shape of spores or germs. His experiment was a perfect success. Sterilized fluids kept just as perfectly when sealed with a wad of cotton, as when sealed hermetically. This, as it became established, effectually disposed of the theory of gaseous ferments. But the claim was now made that the dried remains of decomposing bodies might fly about in the air and renew their molecular motions on being again moistened, and that these would be kept out of sterilized fluids by the cotton batting, the same as was claimed for living germs.

The following are the conclusions of Schroeder :—

1st. "All vegetable or animal forms derive their origin from other living vegetable or animal beings. Omne vivum ex vivo.

2d. "When a series of specific products of fermentation and putrefaction are developed at a certain spot, germs which originate the process have been conveyed to that spot through the medium of the air.' Such is certainly always the case with regard to germs of mould, and to the ferments of wine, milk and urine.

3d. "Vegetable and animal matter in which all germs have been destroyed by boiling, and which, while yet in a hot state, has been shut off from the direct influence of the external air by means of cotton-wool, remains perfectly free from mould, fermentation or putrefaction. The germs, which otherwise would be supplied by the air, are arrested in the passage of the latter through the cotton-wool.

4th. "The germs of most vegetable and animal substances are destroyed by exposure for a short time to a temperature of 100° C. (212° F.)

5th. "But milk, yelk and meat contain germs which are not thus killed. Boiling at a high temperature, under higher pressure, or long-continued boiling at 100° C., will, however, always suffice to destroy these germs.

6th. "The germs in milk, yelk of eggs and meat, after having been boiled a short time, are still capable of being developed into the specific ferment of putrefaction, and sometimes, also, those of yelk; at least, into long and indolent vibriones.

7th. "The specific ferment of putrefaction is of an animal nature. It develops itself and multiplies at the expense of albuminous compounds, but does not multiply under those conditions alone which supply all the requisites for vegetable growth." (Sydenham Society's Year Book for 1862. Annalen der chemie u pharmacie, vol. 117.)

PASTEUR.

The next important advance may be best illustrated by the labors of M. Pasteur, in his sealed-flask cultivations, although he was not the first to employ this particular mode of study. Shutzenberger mentions that this kind of cultivation was first performed by Needham, in London, who published a work on this subject in 1745, in which it seems that he maintained the hypothesis of the de novo origin of infusoria. Spallanzani, a celebrated Italian physiologist, took the matter up, which resulted in a lengthy controversy, in which Spallanzani refuted, by experiment, the conclusions of Needham. The controversy turned on this point. Spallanzani was not satisfied with heating the *hermetically* sealed flasks containing the infusion for several minutes, merely the time required to cook a hen's egg and destroy the germ, as Needham expressed it, but he kept them boiling for an hour; after which no infusoria were found. (Shutzenberger on Fermentation, page 311.)

The plan of Pasteur was practically this: He reasoned that if he could grow the yeast plant in sealed flasks until all gluten with which it was first *connected* could be certainly got rid of, he could then study the plant in all its purity, and

effectually disprove the molecular motion theory. He, therefore added a small quantity of yeast, as pure as possible, to
some cultivation fluid which had been effectually sterilized by
heat and contained no gluten, and carefully sealed it. After
this had passed through the stage of fermentation, another
flask was prepared in the same way; and under the utmost
precautions that nothing else should be introduced from without, a drop was taken from the first with a pipette, and added
to the second, and so on, to the fiftieth generation, and in many
of his experiments more than this number. These flasks
were, of course, kept at the proper temperature for the development of fermentation. The result of this series of experiments
seemed to satisfy Liebig's basis for the recognition of living
beings; the yeast plant reproducing itself after its own form,
for generation after generation, continually, and after the
fiftieth or the one hundredth generation, again producing its
characteristic effects upon must or wort, namely, vinous fermentation. They did more than this. It is well known that,
under certain conditions of temperature, vinous fermentation
passes into acetic fermentation, with the production of vinegar.
Now, in his investigations, Pasteur found that he was able to
separate these forms of fermentation by special management
of the cultures. By taking at the beginning a fermentation
contaminated by the acetic process and keeping the temperature
high, the acetic acid plant, the microderma, flourished, while
the vinous plant, the torula, gradually died out. He thus
rendered the conditions unfavorable to the one and favorable
to the other; and by this operation and the reverse one he
succeeded, after passing the culture through many generations,
in obtaining the vinous yeast plant entirely pure, and the
acetic yeast plant entirely pure, each of which would always produce its characteristic fermentation and nothing else.
And what is more, he found the plants to be characterized by

different forms, which remained constant. I use this simply as an illustration of the plan of what was known as fractional cultivations. The same was done with the lactic acid fermentation, and its characteristic plant isolated and established. In time and by this process several others have been successfully isolated and established, until all of the processes heretofore known as fermentations have been certainly connected with living organisms, without which none of them can succeed.

Pasteur, after working out vinous and acetic fermentations, and that peculiar fermentation which destroys, in its turn, the vinegar, has followed up the process of the complete return of the fruit to the primitive inorganic state.

After the fermentation of the vinegar, Pasteur next examined the putrid fermentations, or decompositions, as they are usually called. The principal agents in these he found to be certain very active bodies which he called vibrios, and which multiply with remarkable rapidity in the depths of the mass, as much as possible apart from oxygen. This characteristic position serves to distinguish them from the ordinary bacteria, which require oxygen for their growth and are found near the surface.

These vibrios of Pasteur are the proper authorities which administer upon the ruined estates of dead bodies, animal and vegetable.

Countless swarms of bacteria may assist them, by their presence near the surface, by fixing any oxygen which would otherwise be absorbed by the fluid and hinder the work of the vibrios, which work without free oxygen. Pasteur has grown these vibrios in flasks, from which every particle of free oxygen has been removed, and found them to be the true agents of decomposition.

His fluid for this purpose was prepared by placing in a three-liter flask of water the following salts:—

Pure lactate of lime	Grams	75.0
Phosphate of ammonia	"	0.5
Phosphate of potash	"	0.4
Chloride of magnesium	"	0.3
Sulphate of ammonia	"	0.2
Sulphate of soda		a trace.

Pasteur has also cultivated, individually, many other organisms, and noted their peculiarities. The extreme high temperature borne by the spores of the ordinary moulds is very wonderful. For instance, the Penicillium Glaucum grew after being exposed for a considerable time to a dry heat of 248° F. and 257° F. It was the same with the spores of other mucidines. At 266° F. the power of growth was lost in them all.

Spores of the bacterium lactis were destroyed at 250° F.

Pasteur also made many experiments to determine the prevalence of germs in the air. He placed fermentable fluids in flasks, boiled, and sealed them while hot; after they were cold he broke the end of the tube, letting the air enter unmodified, then sealed them again at once. It often happened that no organisms developed in these flasks, but they generally did in Paris. In the open country only eight out of twenty developed germs. On the Jura only five out of twenty. On Mountanvert, in a wind blowing from a glacier, only one out of twenty contained germs.

There was one point of failure in Pasteur's work which it seemed impossible to settle by his plan. In all his experiments the organisms are transferred to his flasks together with a minute portion of the fermenting fluid containing them. Now the claim is made that it is this minute portion of decomposing fluid that sets up and maintains the act of fermentation, instead of the organisms, as claimed by Pasteur. This claim has much influence with all those who are inclined to the views of Baron Liebig, and its force must be admitted.

Pasteur's work made a very marked impression upon the opinions held by the scientific world. Although the general facts as developed had, in a measure, been accepted before, there had not been so clear a demonstration of them. The result was a vast increase of confidence in the germ theory of disease, as held by Schwan and others nearly a quarter of a century before. And when Liebig wrote his last paper on this subject in 1870, reaffirming his old doctrine, with some modifications, it produced but little impression upon the growing faith in the new doctrine, which now numbered among its adherents a large number of the most thoughtful and best informed scientists. (See Half-yearly Compendium, July, 1870, page 38, for Brief of Liebig's article.)

The following statement contains Pasteur's conclusions: " The chemical act of fermentation is essentially a correlative phenomenon of the vital act, beginning and ending with it. . . . There is never any fermentation without there being at the same time multiplication of globules, or the continued consecutive life of globules already formed."

SECOND LECTURE.

BASTIAN.

The presentation of this subject would be incomplete without some reference to the view of the de novo origin of life, "Archebiosis." This view has been held by learned men in all historical ages of the world, though it has never, at any time, been generally adopted. Such a view was, perhaps, more prevalent in the fourteenth and fifteenth centuries than at any other time in the world's history. Bastian, the greatest exponent of that view, however, lives in the present, and I have thought best to defer mention of it to this time, especially as it affects our subject, "The Germ Theory of Disease," only incidentally.

This great experimentalist, in his theories, does not depart markedly from the theory held by the chemists and vitalists, but in a large degree adopts the views of both and forms a *connecting* link between the two. It is, also, here again noteworthy that the facts developed by either and both these are made to serve Dr. Bastian's purpose almost as well as those developed by himself. Taking the molecular movement theory, as developed by the great exponent of that view of fermentation, Liebig, he claims that this same molecular movement, under certain circumstances, actually passes over into vital manifestations, furnishing the living forms found in the culture fluids contained in the experimental flasks of Pasteur.

He claims to have produced these living forms experimentally, especially from infusions of hay and turnips, which he first boiled in his flasks, and while boiling, hermetically sealed,

3   35   .

and afterward exposed to favoring conditions of temperature and light. With these and other infusions treated in this manner, he found within a few days multitudes of living forms.

In order that these shall develop, he says that the flasks should not be more than half or one-third full, and that above the fluid there should be a partial vacuum. When it was urged against his conclusions, that the temperature was not sufficiently high, nor sufficiently prolonged, he increased both, running the temperature to 300° F., but still obtained the same result.

Dr. Bastian enters into a very exhaustive argument, using the facts developed by both the chemists and vitalists, for the maintenance of his theory.

He claims, stoutly, that the molecular motion theory of Liebig is the true one, as far as it goes, but does not go far enough, and quotes largely from him, while he also claims that the fermentation theory of the vitalists is true, from the point where they begin, but they do not begin early enough, and quotes largely from them. In a word, Dr. Bastian claims that these two theories are the opposite ends of a complete cycle of manifestations, with the middle and most important manifestations left out by both parties; i. e., that the one passes over into the other.

He takes the case of the production of vinegar. In the ordinary and natural production of vinegar, the fermentation is always accompanied by the presence of, or as Dr. Bastian claims, the production of, the plant known as the Mycoderma Aceti.

The chemists claim that this plant is an accident of the fermentation, and not necessary to the operation, and succeed by other means in producing vinegar, without the presence of the plant.

The vitalists admit the production of vinegar by purely

chemical means, claiming that this production is not properly by fermentation, and cite the fact that the oil of almonds and a host of other substances naturally vegetable products may be produced by the chemists through purely chemical processes. Dr. Bastian now claims that both processes of the production of vinegar are fermentative, and that the second form (vital) is a more fully developed form of the first (chemical), and that the one, under favoring conditions, will pass over into the other without any introduction of germs.

Other scientists and experimentalists have very generally denied the correctness of Dr. Bastian's results, and set up the claim that he has not used sufficient care in the exclusion of germs, or that the heat to which his infusions were exposed was not sufficiently high, or not repeated sufficiently often, to destroy germs or spores not yet hatched. Moreover, other experimenters have failed to confirm his results. Whatever be the final judgment as to the correctness of Dr. Bastian's views, it is certain that this course of experiments and the discussion which they invoked greatly extended our knowledge of the preserving of fruit in cans, so that we are now able to preserve articles which previously had resisted our efforts. One of these is green corn, which seems to require that it be sealed up at the boiling temperature, and then boiled for a number of hours each day or two, until five or six boilings have been had.

This is accounted for in this way. The spores of certain organisms find green corn a very suitable soil for their development, and while the boiling temperature will destroy the developed organisms it will not affect the spores. Therefore these must be hatched and then destroyed. Hence the intervals of boiling must be continued from time to time, until all spores are hatched and destroyed. After which the corn will keep for any length of time.

LOGICAL SEQUENCE.

We may now turn our attention especially to the germ theory of disease; and in doing so, it will be necessary to go back to the discoveries of Schwan and Latour. It was shown that after the discovery that vinous fermentation was dependent upon the yeast plant, it was at once assumed that all other fermentations, decompositions, miasms and *contagions* were also dependent upon the life force.

This assumption was a logical sequence to the then existing ideas, for it had long been held by the wisdom of the world that all these processes, including contagions and miasms, were of the same order, of the same genus, and were brought about by forces of the same nature. But when this assumption was put forth under the new phase of the matter resulting from the discoveries of Schwan and Latour, it was vehemently denied by a host of scientific men, even among many who accepted the yeast plant as the true explanation of vinous fermentation.

DIFFICULTIES.

This, as we have seen, rendered it necessary that the facts of each process heretofore regarded as fermentation should be proven upon its own merits.

The work was at once begun, and the more ordinary fermentative processes were developed with considerable rapidity, and have finally been completely worked out and definitely settled, as has been shown.

The proof, however, that the ordinary decompositions, contagious and miasms belonged to the same class and were brought about in the same manner, was maintained with but slight success.

The organisms found in these were of a different order and the most of them more minute, and for a long time it was impossible to classify them. Manufacturers of microscopes were urged to make improvements in object glasses, in order that these minute, ever busy objects might be better seen. And the observer was compelled to await the tedious operations of the maker of microscopes before he could proceed satisfactorily with his work, and then usually found that the gain was but slight.

The subject became more and more complex the more it was studied. It was found that in each decomposition there were varieties of organisms. When one person found an organism which he regarded as distinctive of a certain process or disease, and published his results to the world, it was usually quickly shown that this same organism was to be found in widely different situations, entirely disconnecting it, as a cause, with the process or disease claimed.

This kind of disappointment recurred time after time, until many of the best men turned away from the subject with disgust, believing the theory incapable of demonstration.

It was proven beyond question that many, at least, of the bacteria were incapable of producing disease, and the view that these forms were the accidents of the process, merely accompaniments, were mere scavengers and had no causative relation to it, gained wide credence.

There was much reason for this view. It was in harmony with what is seen to be going on around us every day. Animals are found everywhere that do the work of the scavenger. As the buzzards flock to the dead carcass, so the bacteria swarm on the decompositions.

However, a hawk is sometimes found among the buzzards, for the want of the opportunity of attacking living prey, and if the two were diminished in size to the mere specks which represent the bacteria, the work of distinguishing between them would be difficult indeed. In surgery especially, cases were continually occurring which seemed to indicate that the apparent scavengers became the destroyers of living structures, and there was still a large number of workers who believed that the distinctive features of these organisms could and would be developed in the course of time, and they continued the work steadily and firmly.

MODES OF PROPAGATION.

Knowledge of these organisms was gradually extended. They were gradually divided into groups, according to their forms and modes of propagation. They have been found to multiply in three different ways.

1st. By simple budding, as in the yeast plant.

2d. By fission, in which one individual simply divides into two.

3d. By true spores or eggs, developed in their interior, as eggs are developed in the interior of the segments of the tapeworm.

The forms are quite numerous. The round and oval forms are called micrococci, and the short, stem-like forms bacilli; the spiral forms spirilli; while a large number remain with the common name bacteria connected with a descriptive adjective, as bacteria termo.

More recently the name of the disease found to be, or supposed to be, produced by the particular form has been attached, as bacillus anthrax, bacillus tuberculosis, etc.

Many other names have been used, but those given seem now to be taking their places.*

* Any one who wishes to study these forms particularly should have Cohn's Classification, which is the most complete yet made out, and will doubtless serve as the basis of all future efforts in this direction. *Magnin*, in his little book, "Bacteria," gives this classification in very convenient form.

UTILIZING DISCOVERIES.

Schroeder's discovery, in 1854, that the ferments floating in the air could be filtered out by cotton batting, together with the confirmation of Schwan's conclusions by Pasteur, who had continued his fractional cultivations, applying them to the decompositions, led to an effort to make use of these discoveries in the treatment of wounds, although at that time no individual bacterium had been proven to be the cause of disease.

However, microscopic investigation had shown that they were abundant in wound secretions, and especially abundant in those that took on a bad condition.

These facts, together with the favorable course usual with comminuted fractures in which the skin was not broken, as compared with similar fractures with the skin broken, caused many to regard the difference as in some way connected with the bacteria.

Heretofore, this difference in the healing of such wounds had been charged to the admission of the oxygen of the air; which had formerly been supposed to be the active agent in fermentation and decomposition. It had now been shown, experimentally, by Schwan and Latour, and recently confirmed in the most decisive manner by Pasteur, that the fermentations and decompositions were caused by organic germs.

Schroeder had demonstrated that these germs could be filtered out, and the air thus treated would not cause decomposition. The old hypothesis was completely disproved; and it was most natural to the thought of the time to ascribe the difference to the introduction, through the medium of the air, of the germs of bacteria.

Following up this train of thought, Mr. Lister, then of Glasgow, Scotland, was the first to make a determined effort to reduce it to practice in surgery. Antiseptics had already been long known, and Mr. Lister began his operations by thoroughly disinfecting his own hands, all instruments, sponges, and everything that would come in contact with the wound, including the patient's skin in the neighborhood of the part to be operated upon. He then caused the air in which the operation was done to be disinfected by a continuous spray of dilute carbolic acid, until the wound was closed. Finally the wound was so sealed up as to prevent the ingress of germs from without. He found in actual practice that operation wounds thus treated generally healed without *inflammation*, and without one drop of pus; simulating in every respect the healing of subcutaneous wounds. The publishing of these results by Mr. Lister, in 1865, startled the surgeons of the world, and many grave heads shook doubtingly, as the words were pondered.

The new plan, however, was tried, far and wide. It was a matter that every skillful surgeon could do for himself, test for himself, and judge of for himself. Although failures were made, the aggregate of results were so vastly superior to anything heretofore attained, that it completely revolutionized hospital practice throughout Europe.

EXPLANATORY.

This statement requires a word of explanation. The effect was not so much felt in private practice, or in small private institutions. The reason for this difference may be explained in this way. It had long been demonstrated that large public hospitals became unfavorable to operation or accident wounds. In large institutions, all open wounds were prone to take a bad course as compared with similar wounds in private practice. The supposition that such hospitals became infected with this class of disease-producing germs had already been widely entertained; and now the antiseptic method of Lister was found to remove the difficulty, and render the treatment of wounds in such hospitals as effective and safe as the same class of wounds in private practice, and in most cases even more so. This grand achievement of the germ theory lent an immense impulse to the study of the subject. Every great surgeon became a student of fermentative change, and of the influence of living organisms, no matter what his bias on the subject. As might reasonably be expected a corresponding advance has been attained.

Up to that time no one particular organism had been singled out and proven to stand in a causative relation to any one disease.

But now the study of the individual character of the organisms found in wound secretions, and in the tissues immediately after death from particular diseases, sprang to the front, and has been carried forward by the most acute minds of our times.

NATURE OF THE EVIDENCE.

We may now turn our attention to the examination of the evidence upon which these views were based. The theory, so far as it relates to the production of disease, has been founded on fragmentary evidence, not upon demonstration. And it must be said, also, that much of this fragmentary evidence has been of the nature of what lawyers would call circumstantial evidence. The nature of the evidence, as a matter of fact, may be thus stated: It had long been believed that fermentation, decomposition, miasm and contagion were caused by processes similar or identical in their nature. It has been proven that fermentation and decomposition are dependent upon the life and growth of certain microscopic organisms. Therefore, miasmatic and contagious diseases are caused by

microscopic organisms. This kind of evidence has served as
the basis of practice, and been the means of producing improvements in the management of wounds particularly,
greater than the most sanguine theorist could have expected
thirty years ago.

DIRECT EXAMINATION.

It may surprise some when I say that the first definite
demonstration of micro-organisms in the tissues of those dying
of traumatic disease was made by Rindfleisch, in 1866. Reck-
linghausen and Waldeyer demonstrated the same thing about
the same time. Directly afterward, Birsch Hirschfield found,
by extended examinations, that the unhealthfulness of a wound
stood in direct relation to the numbers of spherical bacteria
found in the pus of that wound. He also found that the
blood of pyæmic patients contained bacteria during life.

HIRSCHFIELD.

Dr. Birsch Hirschfield, on examining daily the pus coming
from a wound, found that, with the ushering in of the first
symptoms of pyæmia, the pus showed a corresponding change,
consisting in the presence of micrococci, either in pairs, strings
or zooglœa (masses of micro-organisms, of whatever kind, often
imbedded in a gelatinous mass, the latter especially when
pyæmia was far advanced or rapid in its course), and in an
altered appearance of the pus corpuscles, which were finely
granular, of less definite outline and lustre, and showed
their nuclei very distinctly, without the addition of reagents.

The blood of such patients contained similar micrococci,
and its white corpuscles had undergone a change very similar
to that of the pus corpuscles. He sometimes found that pus
from a pyæmic patient would contain besides these a quantity
of the bacterium termo or bacterium lineola, which are the

common bacteria of putrescent matters, while micrococci, according to Cohn, Klebs and Hirschfield, are not to be considered as the ferment of putrefaction. (American Journal, page 542, Oct. 1873.)

This is the first recognition of a particular form of microorganism in connection with a given condition that has stood against adverse criticism.

EVIDENCE UNSATISFACTORY.

Within a few years a great mass of such evidence as this was accumulated. It will be seen at once that this kind of evidence is fractional and not conclusive. It shows the presence of micro-organisms, but it is in no way conclusive as to the practical effects of these organisms in causing any given disease. It was not shown that the disease could not progress without the organisms.

NATURE OF MIASM AND CONTAGION.

During this time a theory was developed for the explanation of the spread of certain epidemic diseases, which deserves mention, not only for its intrinsic merit, but for the reason that it has, by directing thought into new channels, greatly enlarged our knowledge of the cause of epidemics and increased our means of defence against them. We have heretofore spoken only of miasms and contagions.

A miasm, according to the germ theory of disease, is an organism developed, under certain circumstances, in the soil, decaying vegetation, marshes, etc., of certain localities, which can enter the human body and cause disease, but cannot grow a second time in the body of another person—at least does not pass from the sick to the well—and is, therefore, not a contagion.

A contagion is an organism whose habitat, by nature or

adaptation, is the human body, and the spores of which can pass from one person to another, either by contact or through the air, and cause disease in others.

MIASMATIC CONTAGION.

A third class partakes of the nature of both, but differs from both.   It is supposed to be an organism which has one period of development in the human body and another period without the human body, and that these two stages are required for its full development.   Therefore, such diseases are not contagious; they cannot pass from one to another until they have found a suitable soil for the second stage of development and completed their spores.   When this has been accomplished, they are again ready for the production of the disease in man, and not before.

In this way whole regions of country become infected, and persons are struck down with the disease without having seen one sick of it, or having been very near them.   Cholera, Yellow Fever, Typhoid Fever, and various other diseases belong to this class.   They are called miasmatic contagious diseases.

This theory is not based upon any demonstration yet made, either in the human subject or upon animals, but rests upon the known life history of certain parasites of the vegetable kingdom and circumstantial evidence.

Among the most familiar and best proven of the vegetable parasites which run through this kind of cycle, is the common uredo or rust on wheat.   This parasite requires two separate growths to complete its spores.   Indeed, it seems to have a double set of spores.   The spore formed on wheat will not again grow upon wheat, but will grow on the leaves of certain bushes, especially the barberry bush.   And the spores formed on these bushes will, in turn, grow on wheat, thus completing its cycle of existence.

It may be that the organisms of the miasmatic contagions require free oxygen at a certain stage of existence, especially for the hatching of the spores formed in the body, after which the organism itself may enter the body, and multiplying, cause disease. Some examples of this nature have been observed in organisms inhabiting putrid substances. The hatching and early stages of existence take place on the surface, but the after life, the real activity of the organism, is in the depths of the mass, away from free oxygen. According to this idea the spores passing out from the sick person must first find a suitable soil for hatching and the beginning of development, before they are ready for growth in the human body.

The efforts already made on this theory, for the prevention of the spread of epidemics, have been sufficiently successful to furnish an additional point of circumstantial evidence in favor of its correctness.

INFECTION EXPERIMENTS.

Turning again to the experimental evidence, we find that fragmentary experimentation has taken a new departure. Men have arisen from their microscopes and begun injecting the bacterian fluids collected from wounds, and from those dead of disease, into animals, and watching their effects. In the first of these, the animals seemed to have been killed outright by the amount of poisonous material injected, and nothing was gained.

Cose and Feltz were the first to obtain valuable results. They injected a small quantity of blood from a patient just dead of puerperal fever, into rabbits. The rabbits sickened and died of the disease induced. The blood from these was injected into other rabbits, which also died after a disease of the same nature. This was repeated many times, always with the same results. These observations were repeated and corroborated by a very large number of observers, within a short time, and Davaine claims that after twenty-five successive transmissions, he found one-trillionth part of a drop of blood sufficient to cause the disease and produce death. The disease induced, however, was pyæmia. And in these experiments this remained constant, no matter what disease the patient had from whom the blood was taken.

The developments of this series of experiments were very curious. It was found that if the blood was filtered, and the serum thus obtained injected, sickness and death resulted, but no pyæmic abscesses occurred. The disease induced was different. Now, a long time ago, pyæmia was supposed to be caused by the absorption of pus. This was denied, and after much discussion of the subject, there was an effort to abandon this term, which was supposed to misrepresent the facts. Septicæmia was introduced instead. The new name, however, only partially succeeded in displacing the old; for, when septicæmia occurred, with the formation of abscesses, it was still termed pyæmia, to distinguish this accompaniment. This series of experiments seemed to demonstrate that, in the one case, septicæmia is caused by absorption of the fluids contaminated by the effete material of the cocco-bacteria; while the other, pyæmia, is caused by both the absorption of the fluids and the entrance of the cocco-bacteria.

Recently, however, it has been shown that each variety is caused by a special form of micro-organism. The organism causing septicæmia being so minute as to get through the filter, had escaped the earlier investigators. Dr. Koch of Berlin, and others, have clearly demonstrated that the fluids deprived of bacteria will produce transient poisoning; or if in sufficient quantity, will kill; demonstrating its poisonous nature.

With this study came a much greater confidence in the germ theory on the part of many zealous observers. Yet many held to the views of Liebig, and showed that, with the exception of gaseous ferments, which were certainly disproved, all the facts could be accounted for on the molecular motion, or chemical theory. It was claimed that the dried remains of decomposing bodies might fly about in the air, in the form of fine particles, and renew their effects on being moistened, as claimed for organic germs, and be kept out of wounds, or be destroyed by the same means; that while micro-organisms were accompaniments, they were not a cause of disease. We will give the views held by some of the great men of the time, as showing the drift of the thought and the entangling of views that have occurred.

THIERSCH.

Professor Thiersch, writing in 1875, after alluding to the success of the Lister method of treating wounds, and sketching the history which led to its adoption, seems at that time not to have been fully convinced of the truth of the germ theory of the origin of septic influences. Granting the full force of Schroeder's discovery, he thinks it still possible that the power of generating putrefaction may exist in the fluid product of putrefaction independently of organisms; that this power may be retained in the dry remains of such fluids, which float in the atmosphere, and be filtered out by cotton, the same as claimed for organic germs. Consequently, he then considered the question an open one, as to whether or not putrefaction can progress without organic germs. He also objects to the deductions from these experiments, because albuminous substances are always changed by heat in the experimental flasks. The argument, or claim, that bacteria are secondary to chemical changes, had been very widely held, and it was not without strong grounds in its favor. In Pasteur's flasks the organisms were always transferred to fresh flasks, together with a small quantity of the fermented fluid, which, it was claimed, prepared the way for the development of the bacteria.

LISTER.

Mr. Lister, at the London Congress, Vol. 2, page 371-2, relates that he drew blood, with antiseptic precautions, from a vein of an ox, into small, purified bottles, and allowed it to clot. He then introduced various quantities of ordinary tap-water (London hydrant water) into the different bottles. He was surprised to find that even eight or ten drops failed to set up putrefaction in the blood serum, while the one-hundredth part of a drop of the same water was always sufficient to set up putrefaction in milk. He experimented with the blood of various animals with the same results.

He found, however, that if the blood serum was diluted with water purified by boiling, the smallest portion of tap-water would then set up decomposition. He also found that putrefying blood, very largely diluted with water, purified by boiling, did not readily set up putrefaction, while the smallest possible amount of the undiluted putrefying blood set up the process at once. Speaking of the result, he says : "How it is that *diffusion* of the bacteria renders them incapable of developing in the serum, I do not profess to understand. It may, perhaps, be, that when the bacteria are introduced directly from putrid blood, the *products of the putrid fermentation adhering to them* may induce, chemically, an alteration in the normal quality of the serum, which, when thus impaired, *may prove amenable to the nutritive energies of the micro-organisms*; while, conversely, copious ablution with water may remove from them the associated substances which may thus act as their pioneers. The view might be otherwise expressed by saying that the bacteria, per se, are unable to grow in normal serum, and can only develop in the liquid when vitiated, whether by the addition of water or by the action of small quantities of the acrid products of decomposition.

"Or, again, it seems to me conceivable that the normal serum may oppose an insuperable obstacle to the nutritive attractions of an individual bacterium, but that this may be overcome by the associated attractions of several of the organisms in close proximity, after the analogy of the more energetic operation of a concentrated solution of a chemical reagent. But whatever be the explanation the fact remains."

I give the first statement of Mr. Lister as the strongest statement of the view against the primary causative effects of bacteria that I have seen, while I regard the second as approaching very nearly a true explanation of the facts in the case. I will refer to this again.

One fact Mr. Lister thinks demonstrated beyond question. Pure blood resists decomposition under circumstances in which very many other substances decompose, and that a blood clot is a much better antiseptic dressing than was supposed. Most forms of bacteria must be grouped together in some numbers before they can live and grow in blood; while a single bacterium lactis detached from others and washed with purified water will be sufficient to start decomposition in milk.

DR. SALMON.

The idea that pathogenic bacteria must be grouped together in considerable numbers in order to grow in the blood or tissue of animals, has recently been strongly confirmed by the experiments of Dr. Salmon (see Medical Record, April 7th, 1883), in his experiments with "dilutions" of the organisms

of chicken cholera. This experimenter finds that dilutions
of the fluid containing these organisms first produce a mild
type of the disease which they induce, and finally, as the
dilution is increased, fail entirely to produce an effect, although
there may still remain in the drops injected a number of the
organisms. He believes the vital energies of the animal
resist the growth of the organisms, and that numbers are
required to overcome this. We will allude to this again.

BILLROTH.

Billroth, in the course of his inquiries in the early part of
the sixth decade, published in 1874, came to the conclusion
that organisms were present in the human body and in the
bodies of animals in a state of health. And he supposed
that in case of injury they need not be introduced from without in order to develop in a wound and produce sepsis, but
that they may come from the tissues beneath, and if the conditions of the wound be not rendered unfavorable for them,
they may develop sepsis. Professor Thiersch seems to have
attached great importance to this view.

In the same publication, Professor Billroth held, also, the
view that bacteria were the accompaniments of disease, but
denied that they were a cause. His view of contagion seems
to have been divided somewhat between those of Liebig and
those of Beale (vide infra), that is, he seems to have regarded
both these as operative causes or modes of formation of contagious matters.

In his introduction to a report on surgery, in 1876, he takes
occasion to speak of this subject again, and seems to have very
materially modified his views, as will be seen by the following
sentence, which I quote: "I still hesitate to accept uncon-
ditionally the assurance of our best observers, that zymotic
germs have much to do with the causation of erysipelas,

diphtheritis, septicæmia and pyæmia, but in hospital gangrene there appears to be no doubt that the disease is so caused."

Since that time Billroth has fully recognized the demonstrations of Dr. Koch. (See Belfield's Lectures.)

PROF. THIERSCH.

Prof. Thiersch relates the following accidental infection experiment: "At the end of March, 1871, the recently finished St. Jacob's Hospital was brought into use. Some of the furniture from the old hospital (which was not in use because it was infected with hospital gangrene), was piled up in one of the courts. After some weeks it had disappeared, and the author was informed that it had been removed to a more fitting place. Two days later, a virulent type of hospital gangrene appeared in two wards (of the new hospital), remote from each other. A search for the cause of this gangrene, which had not been introduced from without, exposed the fact that the old infected furniture had been stowed in two basement rooms, from which these two wards drew their fresh air."

VIRCHOW.

Prof. Virchow speaking on this subject at the International Congress at London, especially of the claim so often made that micro-organisms could only exist in diseased organs, that they were scavengers, which only attacked the dead or dying, said he thought the view was one-sided. It was necessary to distinguish between different forms of these organisms. Some very virulent organisms can multiply anywhere, in all tissues and at all times.

"The bacillus anthracis had no need of diseased tissues. It is vigorous enough to destroy any tissue, and will develop itself in a short time, from the point of introduction, into a

large focus of disease. So with diphtheria. It is not necessary for any diseased part to exist before the disease com-
mences; that is, the organism finds its way into healthy
parts."

" The action, then, of the virulent organisms is very different, and it is more philosophical to explain the facts in this
way, than to refer the differences to .the mode of action of
the parts."

" Part of them (kinds) will penetrate into the interior of
cells, and there multiply, while the cells will perish, destroyed
by the evolution of the parasitic organism."

In other cases (kinds) the organisms pass between the cell
elements into the interstices of the tissues, or into the inter-
cellular substance. In this case the action of the organisms
upon the cells is not direct. *The cells are only affected by
some matter secreted by the organisms,* or by some chemical
substance produced by the action of the latter upon the fluids
of the body.

PROF. KLEBS, OF PRAGUE.

Prof. Klebs, of Prague, a worker in this field, of great
ability, in his address upon this subject at the London
Medical Congress, discussed the views of Bastian, of the
spontaneous generation of organisms; and in summing up,
says that " since it is abundantly shown, by observation which
is trustworthy and reliable, that the material cause of disease
is introduced from without, therefore, the subject of spontaneous generation has no weight in this discussion."

" Those organisms produced in the body, or those normally
inhabiting healthy tissues, if such were proven to exist, could
not be regarded as a cause of disease. Only one possibility
can here be entertained. Can the organisms ordinarily existing in the healthy body (if such there be, which Klebs denies)

be so changed, on account of change in the fluids which they inhabit, that they may become a cause of disease. No disease has been shown to arise under conditions which would make this seem probable." Continuing, he says, "We may proceed from this proposition as the basis of future inquiry." "Specific diseases are caused only by specific organisms." "The question now arises, whether specific differences of a morphological nature, which remain constant, can be shown to exist in such organisms, so that, beforehand, their nature can be shown by their forms." "It is possible that the specific effects may be due to fine chemical differences in the working of these organisms." "But, happily, even here the general law of nature holds good, that difference of performance is represented by difference of form, and that within the same form lines there are different grades, only, of performance or capability."

Prof. Klebs thinks that the knowledge already attained is sufficient to demonstrate that the forms of the organisms causing disease do remain constant, and produce the same phenomena and nothing else, under given conditions. He attempts a classification, but admits that our knowledge is not yet sufficient for a satisfactory basis.

VOLKMAN.

Prof. Volkman, of Halle (London Congress, vol. ii, page 362), took strong grounds in favor of the germ theory. He says, "We know of no suppuration which does not depend on the influence of organic ferments. . . . In the worst constitutions, and with the most disordered state of health, no suppuration takes place if septic infection is prevented. . . . If we could completely exclude all organic ferments, the largest open wounds and the most extensive loss of tissue would heal without suppuration."

LISTER'S REPLY.

This was evidently regarded by Mr. Lister as a very extreme view. And he replied, speaking very strongly against the conclusions reached by Prof. Volkman, Klebs, and some others. He says, " But, gentlemen, while I am more than ever convinced of the relations of micro-organisms to disease processes in wounds, I propose to utter, this morning, what seems to me to be a needed note of warning against a tendency to exaggeration in this direction. This exaggeration, if such there be, is largely due to the success of the antiseptic treatment."

After speaking of the certain success of the antiseptic treatment of wounds, and the banishment of septicæmia, pyæmia, hospital gangrene, etc., he says, "Such success proceeding from a mode of treatment designed especially to exclude bacteria, may suggest that all inflammations are due to micro-organisms, and that suppuration, acute or chronic, is always due to similar agencies. Gentlemen, I believe this to be a very exaggerated view of the matter, and a view which may tend to have a serious influence on our practice."

"For example, if we believe inflammations are due only to the invasions of microscopic organisms, to what purpose do we employ counter-irritation."

Mr. Lister then goes on to show that *inflammation* does arise, in very many instances, without the aid of micro-organisms, and from a variety of causes, both in wounds and separate from wounds. That abscesses form, and pus forms, in many instances, under circumstances that exclude bacteria. Also, that strict search has shown the pus, under such circumstances, to be free from bacteria.

THIRD LECTURE.

BEALE.

It will not be out of place to notice here the peculiar views of Dr. Lionel S. Beale.

In a work published in 1870, entitled, "Disease Germs; their Nature and Origin," Dr. Beale holds a view of this subject different from the more generally accepted theory, and differing entirely from both Schwan and Liebig as to the nature of contagions. We should say, however, that this view was not entirely original with Dr. Beale, but he has given the most perfect exposition of it.

In respect to bacteria, he regards them as accidental accompaniments of disease; their home is among dead matter, having nothing to do with the living. Further, they are of no distinct species or form; a spore may produce this or that form, according to its environment, and nothing can be predicated of the action of any particular form. They are scavengers only.

The action of a contagion is not a fermentative process in any respect, either in accordance with Liebig and the chemists, or in accordance with the views of Schwan and Pasteur. All the facts he thinks susceptible of a different explanation.

Dr. Beale regards disease germs as being derived directly from the human tissues by a process of physiological degradation—by a metamorphosis of functional activity—without there being necessarily any change of form from the normal cell or bioplasm. This is the bioplasm theory of contagion. It has also been termed the amœboid theory.

Dr. Beale believes contagious disease germs to consist of

outgrowths from the white corpuscles of the normal blood, excited by germs of a similar nature. This outgrowth occurs in successive changes in the physiological activity of the bioplasm of each successive globule of living matter, as new globules are given off from the parent globule, until the type of the exciter of the change has been reached more or less perfectly. After these changes have been induced, and have run their course, many of these bioplasts die and are eliminated from the system. But very many are eliminated in the living state, and the person or animal, if it has been able to withstand the attack, returns to a state of health.

The germs eliminated in the living state have acquired a wonderful tenacity of life, and are capable of living for a very considerable time in the atmosphere. They have the power of entering into the circulation of another animal by way of mucous membranes and otherwise. And when so entered, they induce precisely the same changes in the blood as before described.

This is an extension to the contagious diseases, with some change of form, of the theory now generally held as to the nature of the formation of neoplasms, or new growths in the form of tumors, such as cancer, sarcoma, etc.

These are regarded as changes in the physiological character of the cells of the part, by which they take on new properties resulting in abnormal proliferation of cells with like characteristics. These cells, when transplanted to other animals, will, if the conditions be favorable for them, grow and manifest all their acquired physiological characters. In this way, epithelial cancer may be transferred from man to the animals, or from person to person, perhaps, as readily as syphilis from person to person, except that the cancer cells must be actually grafted into the second person.

Moreover, it is generally recognized that the infiltration

which occurs so generally in carcinoma is due to the wandering of young cells or bioplasts from the seat of the tumor, which have, through a process of physiological degradation, acquired new properties by which they are enabled to initiate a new focus of disease wherever they may lodge in the tissue. These are found to wander along the course of the lymph streams especially, and to a lesser extent along the blood streams. There is also a certain wandering among the neighboring tissues, infiltration.

Such diseases, however, are never contagious, in the ordinary sense of that term, and such cells do not acquire any power of continued life when isolated from the animal economy.

In arguing the point, Dr. Beale does not insist that in each individual case the white corpuscles of the blood must produce the germs of contagion. "For," he says, "the facts may be accounted for solely by the proliferation and growth of the acquired germs or those that have been implanted from without. But," he says, "the fact that living matter of the blood of one individual will live in the blood vessels of another, that the skin, periosteum and other tissues may be transplanted and grafted, prepare us for the remarkable circumstance, demonstrated by experiment, that living pus bioplasts, which have, indeed, been derived from normal bioplasm, may traverse long distances, free and independent, and then, gaining access to other organisms, may live and grow and multiply in them, and establish changes of the same kind as those which were taking place at the seat of their origin."

In speaking of the first origin of contagions, he says, "It is possible that the exceedingly minute living particles which constitute the ' contagium ' of contagious diseases may be the degraded offspring of some kind of normal living matter or bioplasm, which originally possessed *comparatively* exalted tissue-forming, or other formative powers. . . . . The

4

phenomena which occasion the formation of ordinary pus, may, if they continue to occur for a long period of time, determine the development of a specific pus, which has still more marvelous powers of vitality. So, I think, it may reasonably be argued that if the ordinary feverish state be prolonged for a considerable time, and be severe in degree, it is likely that the bioplasm in the blood, collected in the capillaries, may give origin to bioplasm with marvelously increased power of retaining its vitality and of growing and multiplying. The particles of this making their way through the vessels and escaping may live for a considerable time, and having entered the blood of another person may excite in it the changes which accompanied its own development."

" Finally," Dr. Beale says, " it is not probable that disease germs have sprung from insects, or animalcules, or any kind of vegetable organisms ; neither have they originated in the external world and seized upon man ; but they have been derived by direct descent from the normal living bioplasm of the organism. They have originated in man, and if man is not, indeed, responsible for their origin, he has certainly himself imposed the conditions favorable to their production and dissemination."

It is worthy of note that the experimental proof upon which Dr. Beale relies for his theory is similar to that upon which others have relied for proof that bacteria cause disease, namely, injection by pus and fluids from the tissues, and observations of the transmission of contagions, the means by which they may be prevented, etc. He does not pretend that the character of such contagious bioplasm can be determined by microscopic study, but on the contrary, distinctly asserts that it cannot, and further, that the study of the subject gives no hope whatever of demonstration by this mode of inquiry; for the reason that the bioplasmic forms give no indication whatever as to the character of their physiological activity. As a theory it is very interesting, and is certainly very ingenious, and at the same time quite plausible. It has, however, made but little impression upon the thought of the world, and is now strictly limited to those diseases characterized by new growths of tissue, by a great majority of the prominent thinkers of the profession.

In the discussion of this subject at the International Congress, this theory, as applied to contagious diseases, was mentioned but once. Dr. Richardson, of London, held views almost identical with those of Dr. Beale. (See American Journal of the Medical Sciences for October, 1875, page 516, for Dr. Richardson's views.)

Dr. John Bell holds the same views. (See Half-yearly Compendium, January, 1872, page 112.) Also Prof. Crooks and many other important persons ; indeed, this seems to have been the most prevalent view in England during the eighth decade of this century. Lister's success, however, swerved the thought of Englishmen to the theories of Schwan and Pasteur.

BACILLUS ANTHRAX.

While the experiments detailed in our last lecture were being carried on mostly in Germany, Pasteur was still busy in France with his flask cultivations. Having demonstrated the nature of the ordinary ferments by pure cultivations (fractional cultivations as they came to be called, the development of one species out of many), he turned his attention to contagious diseases, and finally hit upon some brilliant results. The most widely known and best proven of these is the isolation of the bacillus anthrax, the contagium of that dread disease which has made such havoc among sheep and cattle in many parts of Europe, and also has been many times epidemic among men with deadly effect. He obtained this bacillus entirely pure, and succeeded in causing the disease in animals by inoculating them with it, after the fiftieth, on to the one hundred and fiftieth generation, grown pure in his flasks. The whole life history of the organism has been made out, and is now substantially confirmed by a large number of separate observers. Indeed, this bacillus is recognized on sight by observers in this field of work everywhere. It can be bottled up and sent to any part of the world, like vaccine virus, and produces the disease in regions where it was unknown, just as readily as at home. This would seem to be a complete demonstration of the germ theory. Yet strangely enough its acceptance as such has been received with extreme caution. That M. Pasteur had the virus of anthrax, no one could doubt, but was it the bacillus, or was it contained in some other form in the culture medium, and thus carried from flask to flask. It was still claimed that the molecular motion described by Liebig had not been eliminated.

KOCH.

We have now to notice a class of experiments still more conclusive than any preceding ones, and which are as yet unexcelled. But it is first necessary to notice the manner in which they were approached. These are by Dr. Koch, of Berlin, working under the auspices of the Prussian government.

This brilliant experimentalist has been very fortunate in his endeavors to devise more perfect means of recognizing very minute organisms with certainty; and of isolating them by pure cultures. One of his first and most important feats was the devising of a plan of staining these organisms, by which he was enabled to practically isolate them from the animal tissues. By his plan of aniline staining he succeeded in obtaining the organisms colored a dark blue, while the tissues were rendered transparent. So completely was this accomplished, that the animal tissues were practically gotten out of the way of the observation of the organisms. His plans have now been tried by a large number of investigators, and have stood the test of adverse criticism. By this new plan of work it was soon demonstrated that the normal tissues of men and animals are entirely free from these organisms; and that the bodies before thought to be such by Billroth, Thiersch, and many others, were, in fact, the granular constituents of the blood. Billroth and others, after experimenting with the new process, have fully admitted the demonstration. And the point may be considered as fully settled.

Dr. Koch now began his work by the then common plan of injecting septic blood into the tissues of animals. At first, his results were similar to those common with this mode of experimentation. Too much of the material killed, by direct poisoning, while a small amount induced disease. Even the slightest scratch of a blade that had been dipped into the blood of the dead house mouse on which he was experimenting produced the disease and death, in the usual time, upon a second mouse, i. e., from forty to sixty hours. The organism found in the blood of the dead animals was a very minute bacillus, which remained constant and pure. An effort, during these experiments, to inoculate field mice with the bacillus cultivated in this way in the house mice, failed entirely. They proved to be in no wise susceptible to the disease.

Now, Dr. Koch, in his examinations of the tissues of the dead house mouse of the original experiment, had found at the point of injection an organism entirely different from that found in the blood, and which he got rid of in the injections made afterward; having obtained these in as perfect isolation as he could, by the means at that time devised, and tried them on the field mice, he was surprised to find that they produced a different disease entirely, which was not seen in the experiments on the house mouse, for the reason that it was killed by the first and more quickly acting organism, before the second had time to produce its effects. This new disease ran a course identical with the dreaded hospital gangrene. The organism did not pass into the circulation at all. It spread by its growth merely, and destroyed all tissues with which it came in contact, until vital organs were reached and the animal destroyed. The parasite was found only in the diseased tissue, never in the blood.

Dr. Koch says of this: "These appearances lead us to the conclusion that the action of these micrococci in causing the gangrene is somewhat as follows: Introduced by inoculation into living animal tissues, they multiply, and as a part of their vegetative process, they excrete soluble substances which get into the surrounding tissues by diffusion. When greatly concentrated, as in the neighborhood of the micrococci, this product of the organisms has such a deleterious effect on the cells that these perish and finally completely disappear. At a greater distance from the micrococci the poison becomes more diluted and acts less intensely, only producing inflammation and accumulation of lymph corpuscles. Thus it happens that the micrococci are always found in the gangrenous tissue, and that, in extending, they are preceded by a wall of nuclei, which constantly melts down on the side directed towards them, while, on the opposite side it is as constantly renewed by lymph corpuscles deposited afresh."

Inoculation with the juices of the dead tissue never failed to induce the disease. Inoculation with the blood did not induce the disease in any case. When the house mouse was inoculated with this material, the disease induced was precisely the same as in the field mouse. Thus, these two forms of organism, one a very minute bacillus, the bacillus septicus, the other a chain-like cocco-bacterium, were isolated, and afterward obtained entirely pure by culture on dry slides—a plan presently to be described.

Why the field mice were insusceptible to the disease so readily induced in the house mouse, remains a profound mystery. Yet it is in accord with facts that are continually coming to light. It is a well-known fact that certain persons resist the contagium of smallpox, while others around them are struck down; and the same is true of every other contagious disease. It is also a well-known fact that some classes of animals are insusceptible to certain diseases common to other animals nearly related to them. It would seem that the organism inducing the particular disease does not find in this or that animal a suitable habitat, but why, is unknown.

As I have said, this class of experiments had been made by many before Dr. Koch; but he had developed a distinctive feature. He had demonstrated that the same tissues or fluids might contain germs, which, when separated, would produce separate and distinct processes of disease. Heretofore, as has been said, this class of experiments had only succeeded in producing septicæmia and pyæmia, which were regarded as parts of the same process. And this result was induced, no matter what the disease with which the patient from whom the material was obtained, may have died; results very confusing to those who believed that different diseases are induced by distinct and separate germs.

We have already spoken of the fact that these results were very nearly identical with ordinary blood poisoning from wounds. Now in the discussion of Dr. Koch's experiments a new thought was developed. The substances injected were all contaminated with septic germs; and these virulent germs had killed the animals experimented upon, before other and slower acting germs had time to produce their results. The more the matter was discussed, the more certain it seemed that this had been demonstrated. One point of vast import had been attained; and now the experimentation must proceed from the new basis. Inoculation must be with pure virus of some particular disease; not a supposed virus mixed with various germs. Otherwise nothing distinctive could be expected.

As we have said, Pasteur had gained a most happy result from his flask cultures, in the isolation of the bacillus anthrax. But it had cost an immense amount of labor, and shown that the results were very uncertain at best. Besides, there was a serious objection urged against flask cultures, in that small portions of the fermenting fluid were always transferred to the flasks with the organisms. The anti-vitalists claimed that the molecular motion, poison or chemical ferments were carried over by the small portion of fermenting fluid transferred, and that where there were fermentable fluids, this power would be propagated. It is evident that this argument carried with it much force, for such men as Prof. Billroth, Prof. Thiersch, and others of the great medical men of the world, still considered this argument a valid one, which still placed the causative effect of micro-organisms in doubt.

Therefore other means of cultivation were earnestly sought. Dr. Koch, of Berlin, was again the successful man. He devised and brought into successful operation the *Dry Slide Cultivation*, by which the particular organism could be watched during its entire development, by the eye of the experimenter, so that it could be seen, at any stage of the process, whether or not the cultivation was contaminated by other organisms than the one desired.

The plan was this. Suitable material to serve as food, or soil, for the sustenance of the organism was mixed with albumen and sterilized by heat. This was carefully reduced to a jelly that would solidify on cooling, and spread in a thin layer on glass slides suitable for microscopic examination. When the slides were thus prepared, a needle was dipped in the fluid containing the organisms and drawn quickly across the surface of the prepared culture medium, in such a way as to distribute the few that adhered to the needle, along the line of its track. This preparation was then sealed up, or otherwise protected from contamination, and kept at a suitable temperature for the growth of the organisms, and occasionally examined with the microscope. It was found that the organisms grew well on such soil or food; also, that it required the utmost care to prevent contaminations. These, however, could be detected by microscopic examination, and, whenever found, rejected. It was also found that the different organisms present in a fluid could be separated by this method of culture; for in drawing the needle point along the surface of the solidified jelly, the organisms would often be distributed singly, and would develop in little groups that could be seen by the naked eye. The characters of each could be determined by the microscope. These were generally found in the form of little flakes on the surface of the jelly, and each could be detached with a needle and conveyed to another slide, or when desired, to a culture fluid, and kept pure. From these pure growths the organisms are transplanted to other slides, or culture fluids, at will, and for any number of generations, and kept absolutely pure. Inoculations were made with these pure cultivations; and such inoculations were found constant in their effects, whether taken from the fifth or the one hundredth generation.

By this process, it is believed that the difficulty heretofore existing, as to the exclusion of a possible chemical ferment, has been incontestably overcome. The phenomena are watched

4*

by the microscope, and the growth of the organisms is all that is seen. This, when carried on from slide to slide, for several generations, must disconnect the organisms from any possible following of a chemical ferment. So palpable is this demonstration that Billroth, Thiersch, and many others, who had maintained a position of skepticism, have fully acknowledged it.

By use of these pure dry cultivations, Dr. Koch proved his former experiments. He also obtained and cultivated the bacillus anthrax by this mode; the bacillus of erysipelas, the bacillus of leprosy, and some others of less note; and last of all, that is as yet definitely announced, the bacillus of tuberculosis.

RESULTS.

It will be seen that this work is only just begun. While we have circumstantial evidence of the existence of very many disease germs, only a few have been positively demonstrated by isolation and pure cultivations, followed by inoculations and the production of specific forms of disease. These are the

Bacillus anthrax.

Organism of chicken cholera.

Bacillus of septicæmia.

Micrococcus of pyæmia.

Cocco-bacterium of gangrene.

Bacillus of erysipelas.

Bacillus of tuberculosis.

Bacillus of leprosy.

Spirillum Obermeiri, the virus of relapsing fever.

Besides these, there are some others not so certainly demonstrated. A considerable number of organisms are now pretty well known by their continued association with a particular form of disease, but have not yet been demonstrated by pure cultivations and inoculations. The task of isolating one of these forms, to begin with, is generally extremely difficult, for it is often associated with other forms, most of which have no disease-producing power, and the experimenter generally has no guide as to which is the one wanted until he has carried it through the cultivations and has made some inoculations. In many of the diseases he is met with another difficulty, which he often has no guide whatever in surmounting, in the fact that he does not know what animals are susceptible to the disease. A considerable number of diseases in men have not been known among animals, and under existing notions he cannot experiment upon men. And then the work is necessarily very slow. The thought of the present time demands absolute demonstration. No possible error is admissible. Dr. Koch spent two years in cultivating the bacillus tuberculosis and making inoculations with it, before he was satisfied to announce it. He induced the disease by inoculation with this bacillus in over one hundred different animals, watched most of them until death from the disease induced, and examined them macroscopically and microscopically, to be absolutely sure there was no mistake. This is what the scientific world will demand in case of each and every disease. The work will be very slow, and years will be required for its accomplishment.

THE EVIDENCE.

The evidence that disease-producing bacteria are distinct from the bacteria of the decompositions is now becoming very clear. All the decompositions swarm with bacteria of various kinds, that are entirely free from any causative relation to disease, and evidently, may be tolerated in food or in contact with the mucous membranes, and even in wounds, without any injury whatever. Every one who eats butter which has become even a little strong, consumes a great number of organisms, and yet without the least injury. The same is true of many other articles of food. It also seems probable, from many observations, that the results of decomposition, as it proceeds to the destruction of nitrogenous compounds, flesh and the like, destroys most disease germs such bodies may contain at the time of death.

It has been remarked by a considerable number of observers that blood poisoning from dead bodies will occur only before such bodies have proceeded far in the process of decomposition.   It seems that inoculations made with the fluid product of nearly complete decompositions do not produce characteristic blood poisoning.   This is very distinctly asserted by Dr. Lionel S. Beale, who regards it as a strong argument in favor of his hypothesis. He supposes that the bioplasts which cause disease die during the process of decomposition, which, he thinks, could not be claimed in the case of bacteria. However, the fact that one bacterium is often destroyed in the presence of another, probably by the waste product of the second being poisonous to the first, is now pretty well established, and fully accounts for the facts observed.   Alcohol is the proper food of the mycoderma aceti, but the vinous yeast plant will not thrive in acetic acid.

There is much reason, now, to suppose that disease germs are less plentiful in the air than was thought a few years ago. It is shown that a multitude of germs that induce decomposition are not even capable of living in the blood or tissues of men or animals when placed in them. The bacterium lactis, which flourishes in milk and in the animal secretions generally, will not grow in the blood or tissues of an animal.   Also the bacterium termo, which is so widely spread, and common in almost every atmosphere, and is now generally considered to be the cause of the ill-smelling putrefactions, will die in the living tissues or blood of an animal.   Great numbers of organisms grow in the mouth of almost every individual, and we have seen them on the margins of, and in wounds of the mucous membranes, without in any wise interfering with the healing process.

It would be a great mistake, however, to suppose, from these facts, that disease-causing germs are never present in such situations. They may, at any time, mingle with the harmless varieties, and leaving these, enter the neighboring parts or circulation, and produce their legitimate results, namely, sickness of some specific kind.

Disease-producing organisms vary very much in their power of gaining entrance into healthy blood or tissue. The bacillus septicus and the cocco-bacterium of *gangrene*, the most virulent organisms known, are totally unable to enter normal structures. They must enter through some breach in the surface, or await the weakening effect of other disease, to let them in by way of the mucous membranes. But once within the tissues, the strongest must yield to them. Others of a less virulent nature can, of themselves, make the attack successfully without any breach of tissue or weakening effect of other disease. Such are the spirillum of relapsing fever, the organism of diphtheria, and undoubtedly, also, most of those of the contagious diseases. Still others may gain admission without a breach of continuity, but much more readily enter through some such breach, as, for instance, the bacillus of erysipelas.

It will thus be seen that a man, perfect in all his parts, may go among some of the more virulent of these organisms without fear. He may enter the dissecting room and work for weeks, and even years, among the septic germs, dipping his hands freely into the tissues and fluids, so long as his skin is perfect. But a slight scratch upon a finger may let in the septic germs that will speedily cause his death.

SPORES—HABITS.

It seems well demonstrated that a number of the lower forms of life, bacteria, mycoderma, saccharomyces, etc., form their spores under conditions other than those best suited to the active growth of the plant. From a close scrutiny of the manner and time of spore formation of these life forms, it seems that they are formed only under conditions which slowly destroy the activity of the plant. The yeast plant forms spores only after fermentation has ceased and its multiplication by budding hindered. It has been asserted that the bacillus anthracis will form spores in the cultivation flasks only when the temperature has been allowed to fall below that of the blood, and, therefore, cannot form them until eliminated from the animal in which they grow, or after the death of the animal.

A considerable number of organisms, whose natural habitat seems to be fruits, such as grapes, cherries, etc., form spores only when drying occurs; that is, the plant lives on the fruit, multiplying by budding or by fission; but when the fruit gives out and drying up begins, the formation of spores begins also. This seems to be the controlling factor in spore formation in a large number of the bacterial forms. And the philosophy of this act, when we consider it, is rather beautiful, and would seem to be a special fitness of function to needs. Rapid multiplication by the simple plan occurs in the presence of suitable conditions for activity. When these conditions begin to fail comes the organization of the spores destined to preserve vitality under conditions unfit for active growth.

The facts now at our disposal will not warrant the conclusion that this is a law of spore formation for all these organisms. Indeed, we have facts which directly controvert this proposition.

If Dr. Koch has traced the life history of the bacillus tuberculosis aright, it multiplies only by the formation of spores, never by fission or budding, and the spores are formed under the conditions of the activity of the organism. Only this one instance of this exclusive multiplication by spores is recorded, however, in the whole field of research among these forms of life. Many, however, have been found to form their spores during the continuance of the conditions of their activity, as the bacillus septicus, micrococci of gangrene, and some others.

These facts have a very important bearing upon the modes of communication of contagious and infectious diseases, and may finally serve to explain some of their peculiarities. For instance, if it is found that a disease-producing organism produces spores only when there is a lack of moisture, or at a temperature below that of the human blood, it will be clear that the spores will be produced only after elimination of the organisms. If, in connection with this fact, it is found that the developed organism is destroyed by drying, which is inferred, then it becomes clear that infection will not be likely to take place, through the medium of the air, directly from the sick to the well. Drying of the eliminated organisms must first occur, during which the spores are formed, and then the drying must be continued until the particles are ready to be wafted about by the air currents and in this manner be carried to the well persons. Dr. E. D. Salmon (*Medical Record* for April, 1883) has developed the idea that the introduction of the spores of the bacillus anthracis produces a different result from the inoculation with the developed organism, under certain conditions.

If this bacillus be cultivated continuously under conditions which will not allow of spore formation, it will be weakened in virulence, so that its introduction will cause a mild type of the disease. If, however, this weakened bacillus be allowed to form spores, these spores will produce the virulent type of the disease. If this observation be confirmed by further research, it will become an important element in sanitary and prophylactic measures, especially so if it be found true of any considerable number of disease germs.

It will be seen at once that the thought contained here has an important bearing upon the propagation of yellow fever, cholera and typhoid fever. If we suppose that the power of forming spores only comes with more or less gradual drying of the voided organisms, and that these organisms are not eliminated by the lungs and sweat glands but only with the excreta, therefore, not liable to be transplanted by immediate passage through the air, we are at once able to understand the clinical facts observed as to communicability of these diseases.

SUMMING UP.

Having passed in review, briefly, the rise and progress of the Germ Theory of Disease, we may sum up the principal points thus:—

1st. In the seventeenth and eighteenth centuries, intelligent observers of contagious diseases, after much study of the subject, came to the conclusion they were caused and propagated by a process identical with or similar to fermentation and decomposition.

2d. There was much study of the processes of fermentation and decomposition, in order to arrive at a more clear understanding of the causes of epidemic and contagious diseases with the view of prevention and cure. These experiments demonstrated that the fermentations and decompositions were something different from ordinary chemical phenomena.

3d. The yeast plant was discovered by Schwan and Latour, in 1838. These gentlemen announced distinctly that the

chemical changes of vinous fermentation are caused by the
life and growth of this plant. They disproved, by experiment,
the previous hypothesis, that oxygen is the active agent in any
of the similar processes; and, reasoning from this discovery,
came to the conclusion that all the fermentations, decompositions, miasms and contagions were caused by the life force.

4th. These conclusions were attacked by chemists, notably,
by Professor Liebig, who denied both the facts claimed and
the conclusions arrived at, since which time there has been a
continuous discussion of the subject.

5th. In 1854, Schroeder conclusively disproved the existence of gaseous ferments claimed by chemists, by admitting
a ir filtered through cotton batting to sterilized fluids without
causing fermentation.

6th. From 1857 to 1861 Pasteur successfully worked out
all the more ordinary fermentations by his fractional flask
cultivations, and proved each of them to have a specific plant
growth as its cause. He also showed that none of the decompositions could proceed without living organisms, though the
specific organism belonging to each was not clearly made out.

7th. Basing his efforts on the results of the last two, Mr.
Lister introduced his antiseptic treatment of wounds, in 1865,
which has proved a panacea for most of the dreaded infectious wound troubles in all hospital surgical practice.

8th. M. Pasteur, Dr. Koch and others have succeeded in
isolating a number of distinct disease-producing germs, and
causing the specific diseases in animals, regularly, by planting
these germs under the skin, and Dr. Koch, especially, has
succeeded in doing this after freeing the organisms of all
possible following of decomposing matter by growing them
upon dry slides.

9th. This much having been proven, the continued observation of numerous other organisms of distinct form and

character, always associated with specific forms of disease, warrants the inference that these diseases are also caused by specific organisms.

10th. All the foregoing taken together gives strong presumptive evidence that all contagious and infectious diseases are produced by disease germs.

PART SECOND.

FOURTH LECTURE.

THE RELATION OF MICRO-ORGANISMS TO THE PRODUCTION OF DISEASE.

One who has followed closely the experimentation on the relation of micro-organisms to the production of disease, will, I think, accept the following propositions as a basis for a farther study of the subject:—

1st. Experiment has shown that there are many micro-organisms which do not produce disease or injury; that the mere presence of these life forms is not, in itself, a sufficient cause of disease; but that there are certain micro-organisms that uniformly induce disease when they are planted in, or obtain entrance to, the bodies of animals or men.

2d. All through the experimentation that has been had on this subject, it has been noticed that poisons have originated in connection with the disease-producing organisms.

3d. These poisons seem to be a product of the organisms, and, while produced only in connection with the life and growth of the microscopic plants, may be separated from them; and, when so separated, produce direct poisoning similar to that induced by the poisons from the higher plants.

4th. This poisoning is different from the disease that is induced by the organisms, in that it is direct, and comparatively immediate; while infectious disease arises only after a more or less definite stage of incubation.

5th. Therefore, it is probable that each infectious disease is caused by a micro-organism that is capable of developing in the tissues, or blood, and forming poisons through changes wrought in the molecular forms of matter, by virtue of its vital energies; or, the physiological processes of digestion, nutrition and the formation of waste products.

These considerations suggest an extended examination of the physiological phenomena of life, in its varied forms, as related to matter, as the best means of gaining a clear conception of the teachings of the facts developed by experiment. It will be seen by any one who reviews this whole subject closely that, while the experimentation that has been had furnishes the strongest evidence that disease results from the presence of micro-organisms (is brought about in some way through the life and growth of these lowly forms), the *modus operandi* by which they induce disease has not been explained, or such explanation has been but partial.

Indeed, thus far, the principal effort has been to show that micro-organisms are regularly connected with certain processes of disease, and that these in no case occur without this accompaniment, rather than to show how they induce disease. This was necessary, in the beginning of this work, for the collection and collation of sufficient data upon which to construct theories, and found a basis of truth, from which inquiry might proceed to the more perfect unfolding of the mysteries of this important subject. The Germ Theory of Disease, as it stands to-day, is essentially a mass of unexplained facts, developed by direct experiment of the most subtle nature; so subtle, indeed, that there are but few men who have the natural endowments and rigid training necessary to verify them in a manner, and with the certainty, that the thought of the times demands.

Under these circumstances, and with the plain fact before us that the mere presence of the organisms is not a sufficient cause of the diseases that experiment teaches us are induced by them, we naturally turn to the well demonstrated fact that micro-organisms do give rise to poisons, for an explanation of these phenomena. It is our purpose now to make an effort

to trace certain of the phenomena presented by the life force
in its relations to matter, with the end in view of elucidating
this difficult point. In this direction, as it seems to me, we
are most likely to find a satisfactory explanation for some of
the teachings of experiment that have seemed difficult to
understand. In these brief lectures, however, we can only
glance at the principal points necessarily involved in a complete discussion of the subject.

The phenomena that we will now review will be Digestion,
Absorption, Nutrition, and the Formation of Waste Products
in the several forms of life; after which we will study briefly
the relation which the changed forms of matter, produced by
the physiological activity of micro-organisms, may hold to
the production of disease. This is necessarily a biological
study.

DIGESTION.

Digestion is a process of solution and chemical reorganization of food material, by which it is fitted for reception by the
life force, or is brought within the sphere of chemico-vital
influences. The material, when thus prepared, is conveyed
within the organism by absorption or osmosis. Digestion
and absorption are the admitted forerunners of all nutritive
processes—the initial steps by which nutrition is rendered
possible. This is true of all the varied forms in which the
life force is manifest, from the most minute micrococcus to
man. There are substances made use of by the life force
without digestion previous to absorption, but every living
thing is provided with this mode of food preparation.

PROPOSITIONS.

1st. All digestion is accomplished by soluble ferments,
elaborated by the organism, whether this digestion be accomplished in a stomach, alimentary canal, or elsewhere.

2d. Digestion is threefold in its forms: (*a*) Primary. (*b*) Absorptive or Resorptive. (*c*) Nutritive or Assimilative.

Primary digestion is the ordinary method by which food is fitted for reception by the organism. It is accomplished by a soluble ferment, or ferments, secreted by the organism and thrown out (*a*) into a cavity, stomach or intestinal canal, into which the digestible material is brought; (*b*) into the surrounding media in which digestible material is suspended, or upon digestible solids or semi-solids.

Absorptive and resorptive digestion is accomplished among the ordinary tissues, by a soluble ferment elaborated by the tissues, under special circumstances: (*a*) For the removal of tissues no longer needed, as roots of temporary teeth, of bone during changes of form, and of other tissues after their purpose has been served—resorptive digestion. (*b*) For the removal of substances accidentally lodged in the tissues, as animal membrane and catgut ligatures, etc.—absorptive digestion.

Nutritive digestion is performed within the tissues themselves, and by them. It is the act of assimilation, the act by which nutritive material is converted into tissue and formed material.

Primary digestion, in the higher animals, is performed in the alimentary canal. This is accomplished by the aid of certain substances known as soluble ferments. The term *soluble ferment* is used to denote a very different thing from an *organic ferment*. An organic ferment is always composed of living cells, and is therefore insoluble. A soluble ferment performs the act of digestion only, and is not capable of reproduction; while an organic ferment performs all the acts of complete fermentation, and is capable of continuous reproduction. All soluble ferments are products of the activity of cell life, or of the life force in the cell form. In the stomach it is found that certain cells, forming the peptic

glands, secrete a fluid rich in pepsin, which constitutes the
true soluble ferment of the stomach, the office of which is the
conversion of food of certain known qualities into peptones or
chyme. The pancreas elaborates another soluble ferment,
pancreatin, or trypsin, which acts upon another portion of
the food. Other portions of the alimentary tract furnish
other soluble ferments, which in their turn act upon the food.
The combined effect of all these is to fit the food for reception
by the blood, through the process known as osmosis.

It will be seen at once that these processes of digestion are,
in fact, extraneous to the body; that is, they are in receptacles
in which food material is met by appropriate soluble ferments,
by which it is dissolved and its chemical constituents re-ar-
ranged, remoleculized. (I use the word "remoleculize" to
designate that disturbance or change of the molecular form
of matter which occurs under the influence of the life force.
As an example, I may take the change that occurs in starch
in the presence of saliva, which is expressed in the following
formula:—

$$\overset{\text{Starch.}}{C_6H_{10}O_5} + \overset{\text{Water.}}{H_2O} = \overset{\text{Glucose.}}{C_6H_{12}O_6}$$

It will be seen here that the digestive body, ptyaline, does
not enter into combination, but, by its presence, causes a combination of the water and starch, producing a remoleculization
with the formation of glucose. Again, the digestive body
given out by the yeast plant, when in contact with cane sugar,
dissolved in water, induces the following changes:—

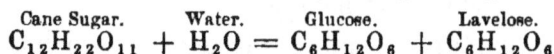

$$\overset{\text{Cane Sugar.}}{C_{12}H_{22}O_{11}} + \overset{\text{Water.}}{H_2O} = \overset{\text{Glucose.}}{C_6H_{12}O_6} + \overset{\text{Lavelose.}}{C_6H_{12}O_6}$$

This result is different in that two new bodies are formed
by the juncture of the two with which we began. In the use
of the word we apply it to any possible molecular change,
without expressing the character of that change. We may
say that the yeast plant destroys sugar by remoleculizing it,

5

or that alcohol is a product of the remoleculization of sugar by the vinous yeast plant, etc.)

It is also sufficiently established that these soluble ferments will act upon food material just as well when removed from the cavities of the body as in them, provided a proper temperature be maintained. This fact shows that soluble ferments, while formed only by the life force, when once formed, act without the life force. A soluble ferment is not itself alive.

FERMENTS.

An example of digestion without a stomach, or the act of digestion of food material suspended in a surrounding medium, is found in vinous fermentation; which is accomplished by the torula, saccharomyces cerevisia, or yeast plant.

The history of the plant is too well known to need description or affirmation, other than the points needed for illustration. When pure vinous yeast is washed with distilled water, a peculiar substance is found dissolved in the water. This is yielded continuously during the life of the plant. Examination has proven this substance to be a soluble ferment, having a peculiar effect upon sugar. This has been examined by M. Berthelot, M. Becamp and others. It has been precipitated and obtained in the form of a powder, somewhat similar to Pepsin, and when redissolved, has been found to retain its original action upon cane sugar. This action is to split up the sugar into two substances, called Glucose and Lavelose, the formulæ of which we have given above.

This reaction always takes place as the primary step in alcoholic fermentation, and is the primary digestion which permits of the appropriation of the food material by the yeast plant. This is entirely analogous to the digestion of food in the stomach of an animal, by which such food is received by the blood to be conveyed to the tissues for their nutrition;

but it is accomplished in the surrounding media instead of in a receptacle provided for the purpose, an alimentary canal. This is one instance of a type of digestion which I believe to be universal in case of all unicellular animals and plants. The formation of a stomach is a provision for the conservation of force, but it in no way changes the *modus operandi* of the digestive function.

Now, if I have been understood thus far, you will probably be enabled to understand the import of certain experiments which have seemed involved in mystery. I will again refer to a series of experiments by Mr. Lister, mentioned in my last lecture, in which he says that he tried a large number of experiments in which he found unmistakably that "it required several drops of tap-water (London hydrant water), in which there were bacteria termo, usually one, to two, or three, in a drop, to start putrefaction in blood serum ; while the least possible portion of putrefying blood (containing, probably, no greater number of bacteria), would start it at once." He wondered at this, and says that "it may be that some substance adheres to the bacteria, or is present in the portion of decomposing blood, that starts the decomposition; whereas, the bacteria washed with pure water are free from this."

In another paragraph, Mr. Lister, always close to the front when searching for truth, strikes much nearer the mark. He says : "Or again, it seems to me conceivable that the normal serum may oppose an insuperable obstacle to the nutritive attractions of an individual bacterium, but that this may be overcome by the associated action of several of the organisms in close proximity, after the analogy of the more energetic operation of a concentrated solution of a chemical reagent. But whatever be the explanation, the fact remains." In this paragraph we find a close statement of what seems to me to be the physiological fact in the case, only Mr. Lister seems not to have reached the point of recognizing the general physiological law appertaining to this subject.

I should explain the phenomena in this wise; the soluble ferment, diastase, or digestive agent given out by this bacterium, must attain to a certain degree of intensity, or concentration, before it is capable of acting on blood serum. A single bacterium is incapable of forming a sufficient quantity of soluble ferment, when dissipated among the serum, to accomplish this digestion, and being incapable of appropriating the undigested material, dies of dyspepsia; while, if a large number are massed together, there is a corresponding concentration of the ferment substance, the digestion of the serum is accomplished, and the colony will flourish. The serum is speedily decomposed.

The soluble ferment of B. termo has been practically demonstrated by Wortman, who has shown that starch may be remoleculized by this organism if no other source of carbon is available. And that in this case a diastase (soluble ferment) is secreted which first transforms the starch into sugar (glucose), which is consumed by the bacteria as fast as formed. If a potato be boiled and a slice cut from it smoothly, to expose the surface, and this surface be exposed to the air for a few minutes, in order that a few germs that may be floating in the air may light upon it, and this be protected afterwards from other germs by a bell-glass covering, with sufficient water to prevent drying, there will be a development of clusters of bacteria lying upon the surface. In ordinary summer temperature this may be seen, usually, after two or three days, as small points of varying colors. Now, if these points be removed, one by one, and examined with the aid of a microscope, it will be found that each little cluster is, as a rule, made up of micro-organisms of one distinct

variety, but the different clusters will usually be of different
varieties. Each of these has evidently developed from a
single germ or spore which has happened to fall at that spot.
By what means do these bacteria appropriate the starch of the
potato to their nutrition? This material is not, as we know,
in a state fitted for the nutrition of these organisms; nor,
even in a condition in which it can be taken up by osmotic
action. The starch must be digested before it can be appropriated. That this digestion really takes place may easily
be shown. If we prepare the potato as above described,
watch it until the micro-organisms have grown to a greater
extent—until they have spread over the surface and the
little clusters have coalesced with each other—then cut a
very thin slice from the surface, and submit this to the chemical test, we will find that a part of the starch has been
changed into glucose, or some form of sugar. This change
is the usual first act in digestion by plants, when starch is
the substance acted upon, and is the primary digestion performed by the diastase, or soluble ferment secreted by these
organisms for the preparation of their food. This change
goes on very rapidly, and the greater part of the sugar formed
is at once appropriated to the nutrition of the organisms, and
its molecular form again changed. In the earlier stages of
the process or growth, it is probable that the sugar formed is
very nearly all used as fast as formed, but when the growth
becomes very luxuriant, there is a superabundant digestion
which permits us to find the traces of the sugar, which reveals
the manner in which the nutrition of the little plants is
brought about.

This process will continue until every particle of the starch
has been remoleculized. But this is only one of the series of
remoleculizations which take place. The starch is converted
into sugar, the sugar into tissue of the growing plants, and

during all this process the plants are giving off waste products. These waste products mingle with the remains of the potato, giving to it its offensive qualities, and it is said to be rotting. We learn from this study that it is not necessary that food material be dissolved in liquid in order that it may be appropriated by bacteria. They are able to furnish a solvent which will bring it into a condition fitting it for osmosis. Otherwise bacteria would be utterly incapable of attacking solids or semi-solids. Yet the cultivation of the pathogenic micro-organisms on solidified media has been rendered popular by Dr. Koch, of Berlin, and it is now fully established that they grow well upon such food. This was described in my last lecture, and is very similar to the growing of the organisms on the potato just described, and illustrates the same physiological laws. Indeed, it seems to have been the observation of the growth on the potato that led Dr. Koch to the dry slide method. By this method, solid food materials may be prepared that are suited to the wants of the particular organism; and, so far as I am informed, any of the micro-organisms may be successfully cultivated by this plan, if sufficient skill is used in the selection and preparation of the material. Under these conditions it could not, reasonably, be expected that the food material could be imbibed by the organisms unless they could furnish some solvent by which it is brought into a condition fitting it for osmotic action.

All these facts point unmistakably to the formation of soluble ferments by the organisms, which serve the purpose, not only of liquefying, but of changing the molecular form of food material; a process in which the analogy to digestion, or conversion of food material into peptones in the stomachs of the higher animals, is perfect.

Many of the micro-organisms seem specially adapted to certain foods, and their digestive solvents are such as to serve the special purpose, as is seen in the ammoniacal fermentation of urine. " M. Musculus has shown that we may obtain from altered urine a soluble ferment. Upon adding to it highly-concentrated alcohol a precipitate is formed which may be filtered and dried. This precipitate, not at all organized, transforms urea into carbonate of ammonia. A temperature of 80° (176° Fah.), destroys it. This diastase appears, then, to be a secretion of the micrococcus urea, and perhaps the role of the bacteria is limited, in the phenomena of fermentation, to the formation of this secretion alone. The ammoniacal transformation of urine would, consequently, enter into the group of fermentations by the varieties of diastase." (Magnin on Bacteria, page 142.)

The proof here given of the production of a true digestive body, or diastase, by the micrococcus urea is sufficiently positive; but M. Musculus must be wrong in his supposition that the " role of the micrococcus is limited to this one function;" for, if the organism does not appropriate from the results of its digestion, wherewithal does it live. It may be that we have here an example of superabundant digestion, which, as a matter of fact, is to be expected when the natural condition is the diffusion of the digestive solvent through masses of fluid material. But the organism must live by remoleculizations within itself, and eliminate waste products. Therefore, the explanation is incomplete.

It has also been shown that the first changes which albumen undergoes in decomposition are similar to digestion. It is first converted into soluble forms, partly peptones, and afterward converted into other forms.

PLANTS.

When we turn our attention to the higher forms of plant life, we are beset with many difficulties. The botanists, it seems, have, as a rule, endeavored to explain the growth of plants from the chemical standpoint, and have almost universally started with the thought that the roots of the plant absorbed water, with whatever might happen to be dissolved in it, and from this the plant appropriated that which it could use in the building of its tissues. This mode of reasoning has, however, failed to give satisfaction. Many facts have appeared from time to time which could not be explained upon this supposition. Prominent among these is the fact that material is continually found in plants which cannot be found in a soluble state in the ground in which the plants grow. Sachs (Text-Book of Botany, p. 625), in discussing this point, says: "But a large portion of food material, especially compounds of ammonia, potassium and phosphoric acid, occur in the ground in a fixed condition, or as it is generally termed, absorbed; they are not extracted from the soil by very large quantities of water; the roots, nevertheless, take them up with ease. It may be supposed, in these cases, that the absorbed food materials occur as an extremely fine coating over the particles of soil, and can therefore only be taken up together with them by the root hairs at the points of contact; and they are there rendered soluble by the carbon dioxide exhaled by the roots. This action of the roots is limited to the points of contact; only those absorbed particles of substance which come directly into contact with the root hairs are dissolved and sucked up. But, since the numbers and length of the roots is very considerable in all growing land plants, and since, also, they are continually lengthening and forming new root hairs, the root system comes gradually in contact with innumerable particles of earth, and can thus take up the necessary quantity of the substance in question. This power of the roots of taking up, by means of the acid sap, which permeates the walls of even their superficial cells, substances which are insoluble in pure water, presents itself in an extremely evident manner, as I was the first to show.

"When polished plates of marble, dolomite, or osteolite (calcium phosphate), are covered with sand, to a depth of a few inches, and seeds are then sown in the sand, the roots, which strike downward, soon meet the polished surface of the mineral, and grow upon and in close contact with it. After a few days an impression of the root system is found corroded in rough lines into the smooth surface; every root has dissolved, at the points of contact, a small portion of the mineral, by means of the acid water which permeates its outer cell walls."

Here we have, clearly set forth, and proven by experiment, a virtual gastric juice, a substance elaborated and sent out to meet food material, dissolve it, and render it fit to be brought in, by osmosis, the same as in the higher animals and in the lower forms of unicellular life.

It is not yet shown, in this case, that this is a specialized soluble ferment substance, but it evidently acts the part of such a substance, and, while this is so clearly shown, its composition makes but little difference, so far as our present purpose is concerned, which is to show that all forms of life are dependent upon digestion in some form, for the preparation of the material upon which they subsist for absorption.

It is not claimed that all material used by the life force must undergo this kind of change. Water is used without such change by all forms of life, and probably many other substances are so used.

SPROUTING OF SEEDS.

With every egg that is formed there is a provision for the nourishment of the embryo until the animal has developed the organs by which it is to prepare its food for assimilation. The egg is composed of two parts, an embryo, and formed material for the nutrition of that embryo. In the seed there

5*

is the same provision precisely. Surrounding the embryo, there is laid up a store of formed material for the nourishment of the young plant until such a time as it shall have developed its roots and leaves, the organs by which it gathers its food, sufficiently for this purpose.

In neither of these is the food material in a condition to be formed into the tissues of the young animal, or plant, without remoleculization. It must be redigested before it can be assimilated by the embryo. This process has been best studied in the seed. Here it is found that the embryo eliminates a true digestive substance, called diastase, the office of which is the digestion of the store of food material. Of this substance Regnault says : " A peculiar nitrogenous substance, called ' diastase,' which possesses the property of converting a large proportion of fecula (starch) into dextrin, and even into sugar, when its action is sufficiently prolonged, exists in the germ of the cerealia and in tubercular *vegetables*. It appears to be formed at the moment of germination, probably at the expense of the albuminous matter contained in the grain, as it resides in the very origin of the germ, and in the eye of the tuber; and its use in the *vegetable* economy appears to be that of disaggregating the amylaceous matter, and transforming it into an isomeric soluble substance, which the vital forces then change into other isomeric, but insoluble substances, such as cellulose, which is to form the framework of the growing plant."

This substance is very well known, and its action on the starch has been closely studied. It may be precipitated by alcohol, and obtained in the form of a powder, and preserved for a long time, and when redissolved in water will still produce its effect on starch. This effect, like that of other soluble ferments, is not to be explained by the ordinary chemical laws. One part of diastase will convert 2000 parts of starch into sugar. The office of this substance, therefore becomes clear; it is the true digestive body of the embryo plant, by which its store of food material is remoleculized and fitted for its assimilation.

In the egg, this process has not been so well made out; but enough is known to determine the fact that the store of albumen is profoundly changed previous to assimilation.

The perennial plant, which lies dormant in winter, is to some extent similar, during the interval of inactivity, to the seed. It contains within its structure formed material for use in spring time, for the reformation of the leaves, for the gathering and transformation of a portion of its food material. These stores of formed material seem to be made use of much in the same manner, and by the same means, by which they are transformed in the seed; that is, by means of a diastase, or soluble ferment, formed at the proper moment for its solution. In this respect, we find a close analogy between the plant and the animal, as we shall see further on.

I have brought forward these well-known phenomena for the purpose of enforcing the principle lying at the very threshold of the nutritive function; for the purpose of impressing the fact that the life force, wherever found, carries forward its business in regular and *definite* forms, by regular and *definite* means; and that its means to the accomplishment of the preparation of its food material is similar throughout all its varied forms. One form of life lives on this, another on that food material, but the process by which food material is brought into a state fitting it for its final assimilation by the life force, and the building of tissue, of whatever kind or quality, is the same in its general plan.

NUTRITION.

The digestion of nutrition is performed by the tissue itself. When no ferment substance is divisible from the tissue the existence of such substance has been inferred from the phenomena observed. We think no one will doubt that the food of the animal goes through further fermentation or digestion after it is taken into the blood, at least, a certain remoleculization, by whatever name it may be called. Is this change brought about in the blood? Posssibly it is, to some extent, especially during its passage through the liver, the great gland of the blood vascular system, the soluble ferment, or diastase, of which is well known. But the great change is brought about by the tissues themselves.

In primary digestion there is simply an absorption of other molecules or atoms, by the molecule of the substance undergoing digestion, or a splitting up into simpler compounds. In the nutritive digestion (nutrition assimilation) there is a reorganization of the elements into new tissue and formed material. This reorganization is performed in the cells themselves. The substances are taken into the interior of the cells by osmosis, and then remolculized by act of the vital force. This is the point, evidently, of the most intimate association of matter with life; it is here that matter is said to be endowed with life.

The products formed here are very definite. The reorganization results in tissue, each tissue selecting and building for itself, and to itself, from the material carried in the blood streams, obtained from the primary digestion and from the lungs. In this reorganization and building by the tissues, definite waste products are given back to the blood streams, namely, carbonic acid and urea. These are cast out by the lungs and kidneys. They are found to be incompatible with life, if retained in any considerable quantity. Here, also, are formed the secretions destined to serve the animal in the performance of the digestive and other functions; in the animal forms mostly by specialized cells acting vicariously for the other tissues of the body. These are special developments of function, seen as we ascend the scale of life. But these functions are also present in the unicellular forms of life, where there is no such division of labor or function developed. It is not a new function, but a special development of cells to act in this matter for the whole.

FIFTH LECTURE.

YEAST PLANT.

For comparison, let us turn again to the yeast plant. Here we have no blood streams to interfere with our study of phenomena. The same cells that emit the soluble ferment which performs the primary digestion, appropriate the results of that digestion, not individually, but en masse. That is, each cell individually excretes, so to speak, the material, the soluble ferment, that performs primary digestion. This digestion is performed by the whole amount of material excreted, independently of the individual cell, but in the appropriation of the results of this primary digestion, each individual cell acts for itself only, and independently of all other cells. The result is the building up of the tissue of the cells by *remoleculizing* the original elements into new forms, with the elimination of waste products.

In this case the waste products are alcohol and carbonic acid. One is the same as in the animal, carbonic acid, the other is different. These waste products agree, in their effects upon the activity of the plant, with the waste products of the animal upon its life force. Alcohol, in a certain quantity, stops the growth, the activity, of the yeast plant which produced it, and if sufficiently concentrated, destroys it. Urea does the same for animal life.

These excretory products differ widely among the lower forms of life, and yet, as we shall show in another lecture, retain a constant resemblance.

In case of the mycodermi aceti, the acetic acid plant, the excretory products are carbonic acid and acetic acid. With

96

the bacteria termo, the excretory products are very complex, consisting of ammonia, sulphuretted hydrogen, and other ill-smelling products of putrefactive fermentation. Some of the mucidines excrete butyric acid; other forms excrete other waste products. Only a few of these have, as yet, been very accurately studied.

Often, perhaps, it is by these excretory products from bacteria that gain access to the blood, and are capable of living there, that disease is produced. One case in point. Dr. Carter, of India, in his very careful and laborious examination of the blood spirillum of relapsing fever, has found that these organisms become still, and diminish in number during the exacerbation of the fever, and during the intervals of comparative freedom from the rigor of the disease they become active and augment in numbers, until the approach of another attack of fever. What does this mean? Certainly the disease is not caused in this case by the primary digestion, by the soluble ferment, as it seems to be in hospital gangrene.

The case seems reversed. Here it seems that the accumulation of the excretory products of the organisms cause the sickness; also, that the amount present in the blood is sufficient to check the development of the organism. Suppose the yeast plant could grow in the blood, and develop its alcohol, might it not intoxicate the man?

The course seems to be this. The soluble ferment excreted by the spirillum affects the patient but little, the development proceeds, and excretory products begin to accumulate more rapidly than they are carried away by the emunctories; finally, when a certain concentration has been reached, the patient is thrown into violent fever. The bacteria are also so affected by their own excreta that they droop and fail to flourish. They become quiet. The patient eliminates the poison, and all is well again with both patient and spirillum, and the repetitions continue. But in this I am anticipating; waste products will be discussed in another lecture.

CELLULAR DIGESTION IN THE HIGHER ANIMALS.

In the higher forms of life we find a division of labor in the primary portions of the nutritive function, the preparation of the material for nutrition. Certain portions of the animal body are set apart, so to speak, to perform this function for the whole. In the lower forms this function is performed by the whole mass of cells as a unit, each doing a part as an element of the mass. In proof of this we have cited the instance of the bacterium, in blood, where digestion can only be accomplished by the united efforts of a mass of cells. We find, then, that in the lower forms of life cells may act for each other in the matter of primary digestion, and that in the higher forms of life a special development occurs which enables certain cells to act for the general mass. But in the matter of nutritive digestion each individual cell must act for itself in the one case, as in the other. Here there can, in the very nature of things, be no vicarious function.

Is it impossible that the cells of the higher animals may excrete digestive fluids; soluble ferments, also, under certain particular circumstances? We shall see.

RESORPTIVE DIGESTION.

We now approach an important point in this discussion, the third form, or resorptive digestion.

We wish to state distinctly that all resorption and absorption of material not primarily fluid, that is, not simply endosmotic, is of the same nature as primary digestion, and is performed in a similar manner and by similar agencies, through a vicarious secretion of a digestive fluid, a soluble ferment, by the proper tissues of the body.

We will take for illustration the familiar process of the resorption of the roots of the temporary teeth. This is a physiological process, as all other resorptive processes are. It is brought about by the action of certain cells which have become known as osteo- or odontoclasts. It has been claimed that the osteoclasts are really osteoblasts, that have changed their function from that of builders of bone to absorbents of bone. The distinction is not important in this lecture. These may be only the ordinary connective tissue cells of the part in immediate proximity with the part to be removed, which have temporarily taken on new functions, the secretion or elaboration of a soluble ferment for the removal of the roots of those teeth which are no longer wanted. These cells perform this act in an indirect manner. It is plain that these roots are not removed by any mere mechanical force. These cells have no physical power of gnawing into them. They secrete a soluble ferment, or analogous body, which digests them, breaks them down, and fits their material for entrance into the blood streams by osmosis, just as solid ingesta in the stomach is broken down and fitted to enter the circulation by osmosis. There seems to be no foundation for the notion that the re- sorbed product of these roots may not form proper pabulum for the building up of other tissues; that the absorbed product is necessarily excreted. Resorption is not a process for the forming of waste products.

As this process has been pretty closely studied, we will examine it a little further. From decay and other accident the temporary teeth are very liable to lose their pulps, which very often, we may say generally, results in the formation of alveolar abscess at the apex of their roots. If such an abscess exists at the time the resorption should take place, such resorption fails, partially or entirely. The tissue which should perform this function is thrown into a pathological state, and the secretion of the soluble ferment does not take place in the normal manner. If, however, the abscess ·does not occur, or if such abscess be cured, and the tissue in immediate proximity to such devitalized root be perfectly healthy, it is found that the resorption of the root goes on in a normal manner. The mere death of the root does not interfere with resorption, provided a physiological condition is maintained in the immediately surrounding cells, which are the active agents in the work. The soluble ferment does not depend for its action upon the life of the tissue to be acted upon. The roots of the teeth are simply digested, and enter the blood streams by osmosis, as any other digested material. Nothing more, nor less.

DIGESTION OF BLOOD CLOT.

What happens when an injury occurs, and a blood clot is formed within the tissues? Is such a blood clot a living substance? Certainly that cannot be maintained. But it becomes organized. How? Certainly the blood clot, as such, does not itself become living tissue.

The process occurring here, by which the clot is removed, is similar in kind to that of digestion—is digestion. All around it, the connective tissue cells begin to throw out a soluble ferment, by which the substance of the clot is dissolved and begins to pass into the blood streams by osmosis. Young cells wander out into the *liquefying* clot in all directions, grow, and furnish their quota of fresh increment of ferment substance to *liquefy* the clot. The question as to whether any of the blood cells (white blood corpuscles) become connective tissue cells under these conditions, need not be argued pro nor con here, for it does not affect our present purpose. Certain it is that cells are found growing out into the clot, and acting very like some of the forms of bacteria. Thus, the clot is digested and removed, and when it is gone any surplus new tissue that may have been formed during the activity of the process suffers the same fate. It is, in turn, digested and removed.

DIGESTION OF LIGATURES.

Of late it has become quite the fashion, among surgeons, to use ligatures, for deep sutures, made of digestible material, such as catgut, animal membrane, etc. These are left in, and it is said they are absorbed. They must be dissolved first. They do not dissolve in ordinary liquids, nor do they dissolve in pus, or upon pus-forming surfaces, or in the usual plasma of the tissues. The surgeon is very careful that they be so prepared that they shall not. He is careful that they are not easy of digestion. The " International Encyclopedia of Surgery" says: "The spontaneous solution of catgut ligatures, when set in wounds, is caused, not by any chemical solution of their structure, nor by any process of organization which they undergo, but by the *invasion of leucocytes*, under the operation of which they vanish, while new tissue takes their place; but, if they be over-prepared (*i. e.*, rendered too difficult of digestion), this change does not occur, and they act like foreign bodies in general."

Here we see that the ligature does not dissolve in the ordinary plasma of the tissue, if properly prepared. The tissue must approach closely a physiological condition, it must throw out certain cells which invade it, as it would be invaded by bacteria if they had the opportunity. These cells must throw out their special ferment for the occasion, to digest the catgut and make room for themselves, as would be done by bacteria; and thus they continue to dissolve out the foreign digestible substance and take its place, stimulated to this action by its

presence. If, however, the substance is too difficult of digestion, over-stimulation results, and instead of the soluble ferment being formed, the cells degenerate, pus forms, and the
process fails.

DIGESTION OF SPONGE.

Let us examine the operation of the sponge graft. In the
sponge, as prepared for this operation, we have a material
somewhat difficult of digestion, a material that would be very
slowly acted upon by the gastric juice. From this circumstance, and the fact of its very minute porosity, it has, happily, been selected for the purpose of inducing granulations
for the filling out of lost tissues. What may have led to this
selection I am not informed. It was probably directed by
some accidental observation. The fact, however, is now well
established, that the tissues are stimulated, by its presence, to
fill its pores with granulations, after which the sponge itself
gradually disappears, leaving the opening filled with new
tissue.

The physiological phenomena may be thus explained: as
has been said, the tissue is stimulated by the presence of the
sponge, and granulation proceeds actively. At the same time
a soluble ferment is thrown out, by which those portions of
sponge inclosed in the matrix of granulations are slowly digested and pass off by osmosis. At the same time, the granulations crowd into the space gained. This process continues
regularly, until every particle of the sponge is removed, and
its place supplied by living tissue. I think it will be clear to
every one, especially to every surgeon, that this sponge does
not dissolve in the ordinary plasma of the blood, or tissues.
The tissues are stimulated to an extraordinary proliferation of
cells by its presence. They are also stimulated to the secretion of a soluble ferment, suitable for its digestion and removal.

The same process precisely is called out in the absorption of the catgut ligature, which is progressively invaded by leucocytes, which digest the foreign substance and remove it by osmosis, making room for themselves, much as would be done by bacteria, and when the ligature is gone we have a cord of new living tissue in its stead. Even ivory driven into the flesh is eaten into, digested, and portions removed, by the invasion of leucocytes.

It will be noticed here, as everywhere, when this kind of digestion occurs, the tissue is applied directly to the substance to be digested, and invades its substance as digestion makes room for its growth, so that the tissue is continuously in intimate relation with it, progressively invades it. We do not suppose that this soluble ferment is pepsin, but a body peculiar to the connective tissues, formed only in connection with a foreign substance.

We will suggest that the sponge graft offers the most feasible means of isolating this peculiar ferment. If several sponge grafts were placed in suitable animals, as dogs, and the animals killed at a proper time, and the grafts immediately treated in some way similar to that employed for obtaining pepsin from the gastric mucous membrane of the pig, this ferment would doubtless be found. The process would perhaps require to be varied somewhat to suit the different character of the soluble ferment. If this soluble ferment could be had in sufficient amount and purity, it would be of some value in testing the different substances suggested for ligatures, etc., and possibly in ways not now thought of. But the principal use of the proceeding would be the demonstration of a physiological process, and the extension of exact knowledge. Such a demonstration would furnish a new basis of fact from which thought might radiate to the unfolding of still other facts, no less important.

NECROSED BONE.

Take the case of necrosed bone. How is a sequestrum formed, but by the digestion and removal of that portion which connects it with the living bone. Can this be accomplished by any of the more ordinary fluids found in the tissues? certainly not. Then a special menstruum is formed for the purpose of this solution. Formed where, and by what? Not in some distant part and sent here, but by the tissues in immediate contact, acting by virtue of their natural endowments and the stimulus of juxtaposition with an undesirable substance. If the tissue could act without undue irritation, the act of digestion of the part should occur, and its place be supplied, without the formation of pus. However, this probably never occurs, except in cases of the death of very small portions of bone. In some cases, however, we believe it does occur to a considerable extent. We have witnessed cases of healing of wounds involving bone, which we can explain in no other way. In most cases, however, the inflammation runs so high, and continues so long, that pus forms, and prevents the physiological action of the cells. These cells cannot perform this function from a distance. The ferment substance, if formed, is dissipated in the fluids; they must be in juxtaposition with the substance to be digested, or at least, no dissipating menstruum must intervene, in order that the integrity of the soluble ferment may be preserved and applied directly to its work.

If this condition can be attained, dead bone will be digested and removed. As a matter of fact, we generally find sequestra eaten into, here and there, sometimes more, sometimes less. These burrowings represent points where the granulations, in a state closely approximating healthy action, have come in contact with and invaded the dead bone. If this condition could be maintained at all points, the sequestrum would be completely removed.

Krause (Algemeine U. Microscopische Anatomie, S. 74), examines the osteoclasts very closely, and the manner in which living bone is removed by them. He is of the opinion that they furnish a secretion for this purpose, and thinks it contains lactic acid. He says, ivory driven into the flesh is absorbed by them, and that the osteoclasts are found in the pits formed in it, as in living bone that is undergoing this process; and quotes Kolliker, Billroth, De Morgan and Tomes, as authority. These osteoclasts must be formed from the ordinary connective tissue cells, or leucocytes. Not from osteoblasts.

ACTION OF TISSUES.

Absorption under pressure is in no wise different from any other process of absorption. A certain portion of the cells, crowded, but not too much, take upon themselves a new action and develop a ferment which destroys their neighbors less favorably placed. It is well known that under pressure, certain tissues are destroyed in preference to others. This would be a very interesting field of research, but we have not time to enter it in this lecture.

This we conceive to be the normal action of the tissues of the higher animals. How do the lower forms of life differ from these? In the *vegetative* sense, the difference is but slight. They differ widely in formative power. In the very low forms the cells fall apart instead of aggregating and unifying into more complex compound forms. In this sense there is a wide physiological difference, but in the matter of digestion, nutrition, and the formation of waste products, the similarity is very close.

SOLUBLE FERMENTS.

We have said enough to indicate the mode of action of bacteria in destroying substances in general, and in causing disease. Some forms may possibly cause disease by their presence, by aggregating into groups and causing irritation, like the animal parasites, as the itch mite; but we are persuaded that their general mode of action is through their soluble ferment and the toxic properties of their waste products, to be examined presently. In this manner the hardest substances are made to yield; are melted down almost as readily as softer materials. Soluble ferments do not depend for their dissolving power upon either acidity or alkalinity. Some are acid, but generally these reactions are not very marked, and normally seem capable of considerable variation. They seem to be Nature's solvents, manufactured by the life force, and used for the maintenance of the creature. We can neither form them, nor tell why they should have such power. We can only examine them, learn their powers, and wonder at Nature's handiwork.

PATHOLOGICAL FORMATION OF FERMENTS.

We have reason to believe, at least to suspect, that soluble ferments are often formed in places where they are not wanted. Tissue is stimulated to false secretion by various irritants and in many ways. It is probable that many of the excoriating secretions which we see about the ears, necks, and other parts, in children, are soluble ferments given out through some malcondition of the tissues. In this manner much harm is accomplished. We will allude to this again.

We have reason to suspect that in some cases decay of the teeth may come about in this way. We have already seen that tooth substance is digested and removed by a soluble ferment normally formed. Why is not the tissue forming a soluble ferment dissolved by it? The answer is this: Any tissue forming a soluble ferment is protected from the action of such ferment substance by the life force. This answer is very clear, and is undoubtedly a general fact. But what is life force? Here we are more or less bewildered, for the want of more exact knowledge. It is well known that the mucous membrane of the stomach of a dead dog is readily digested by the gastric juice of a living dog. Not only this; but the gastric juice in the stomach will, after death, digest the mucous membrane which formed it. In this direction there is a very perfect chain of fact, fully establishing the protective power of the life force.

These ferments are, however, constantly brought in contact with tissues which do not form them, such as the ducts of the glands and other surfaces over which they may be spread. This fact might lead to an extension of our answer. Might lead us to say that all living tissues are protected from the solvent action of the soluble ferments.

This statement, however, goes too far. Certainly it cannot be maintained that the roots of the temporary teeth must die before resorption takes place.

Again; the tissues forming the gums and mucous membrane in advance of the erupting tooth are removed by this same process. It cannot be claimed that they are dead tissues. Very many other instances of the same nature might be cited. Then it is not true that all living tissues are protected from this digestion by the vital principle. Hence, we are driven back upon our first statement, i. e., that any tissue forming a soluble ferment is protected from the action of such ferment substance by the life force. Other tissues are not so protected. We may infer that the tissue forming the lining membranes of the ducts of glands, and any normal receptacle of such a fluid, would be protected from its action, just as the tissue

6

forming the secretion. But it is plain that other tissues are
not so protected, or, if so protected, that protection is in some
way withdrawn, under special circumstances. How? The
nerve force suggests itself; but this hypothesis presents serious
difficulties. It seems much more probable that the explanation is to be found in the development of the power of forming the soluble ferment, than in the withdrawal of the power
of resisting it. In all cases in which I have personally made
examination of the tissue advanced against resorbed surfaces,
surfaces being removed, I have found a marked change in the
general appearance of the tissue. It has taken on the appearance, more or less, of granulating tissue, and these changed
cells have received the name of osteoclasts, odontoclasts, etc.,
according to their location and purpose. The fact that this
peculiar development is the accompaniment of resorption of
bone, roots of teeth, etc., is universally recognized, I believe,
by histologists. The inference, then, that this solvent-form-
ing power is a special development, would seem to be maintained. We conclude that this special soluble ferment is only
formed under special circumstances. Physiologically, only
in case some tissue not wanted longer is to be removed ; as
bone during growth, and consequent change of form, roots of
temporary teeth at shedding time, or when some foreign
substance has been lodged in the tissue. Pathologically,
through some irritation or other abnormal condition, the tissue
is induced to a false action. We can understand that the presence of a substance, such as a catgut ligature or a sponge,
should induce some special action of the tissue, through which
it would be freed from its presence. But how the same
phenomena should be called into play for the removal of
parts of bone during growth, for the purpose of rendering
them symmetrical and perfecting their form, is more difficult
of comprehension. Take, for instance, the phenomena of the

growth of the lower jaw. In the child of five years its body is short, just long enough to accommodate the ten deciduous teeth. A lengthening of the body of the bone occurs, to make room for three large permanent molars on each side. In this lengthening the ramus of the jaw is carried back about one and a half inches, which, in effect, protrudes the lower jaw forward. Now this is not an interstitial growth of the bone ; but the ramus is resorbed away from its anterior aspect and built up posteriorly. The osteoclast cells, soluble ferment-forming cells, are found there, and their effects upon the bone may be seen. Their work, in this instance, however, is done quite evenly and smoothly. This is the type of all changes of form in the osseous system. It will be seen at once that the amount of this work done during the growth, formative stage, of the osseous system is simply immense. The head of each bone must be trimmed down to the size of the shaft as the bone lengthens, etc.

By what power is this action controlled? Here we come directly upon the unknown. If we say it is through the control of the nervous system, we will be brought face to face with the symmetrical formation of bones in the fœtus, in which the formation of a nervous system has failed, in which the spinal canal, or groove, has never closed, is open from end to end, and no trace of nerves is to be found. Yet, such a fœtus has, with the exception of the spinal column and skull-bones, developed to term, a perfectly symmetrical skeleton. After seeing this, the idea that these processes are under the control of the nervous system seems untenable. And I am free to say that I have no hypothesis to advance, except to say that they are under the control of the life force, which, I confess, is no explanation.

PARASITES.

After what I have said, the examples I have given, it seems plain that the work of these lowly forms of life is to cause a remoleculization of the substances with which they come in contact.

This power of remoleculization of certain forms of matter is one of the especial physical attributes of the life force. It is the one great power upon which life depends for its continued existence. No form of life can continue to exist without itself exhibiting this power, or borrowing directly from a form of life that does exhibit it in a marked degree. (It is claimed that certain forms of *parasitic* plants draw from and appropriate to themselves food material which has been assimilated by their host. Even here it is most probable that there is a form of *remoleculization.*)

Now, in case of parasitic plants, if this remoleculization is of such a nature that the results are poisonous to the host, disease follows. If, on the other hand, the results of this remoleculization should not be poisonous to the host, the only harm resulting would be merely from the great crowd of the *parasitic* forms.

The history of the experiment and observation upon this subject, leaves no reasonable room to doubt that remoleculizations occur which are entirely harmless, and that remoleculizations occur with the production of the most deadly poisons. While every imaginable grade between these two may occur.

SIXTH LECTURE.

WASTE PRODUCTS, CONSIDERED WITH REFERENCE TO THE GERM THEORY OF DISEASE.

In the beginning of the germ theory controversy, Schwan seems to have recognized that carbonic acid and alcohol were the excrementitous products of the yeast plant (Annalen der Pharmacie, Band. 29, S. 93 und 100). Liebig, in opposing this theory, *alludes* to this and *recognizes* that this conclusion would necessarily follow, if fermentation were proven to be the result of life force. (Agricultural Chemistry, page 124.) Since this time, however, this idea seems to have been lost sight of by writers on this subject. Sometimes we see indefinite allusions to it; but there is no definite expression of the general law, that all forms of life must have their specific waste products, as they are seen in the animal forms. The remoleculization of matter is continuous with the duration of life. The life force is dependent upon the remoleculization of matter for its support. No manifestation of the life force, in any form whatever, can be conceived of without this accompaniment. As the steam engine is dependent upon fire for its power, so is the life force dependent upon molecular changes in matter for its continuous existence. As steam is dependent upon heat for its generation, and the expansive force by which the engine is driven, so is matter dependent upon the life force for the changes of molecular form through which that life is cognizable to our senses. We can have no conception of life in the material form without the accompaniment of these changes. Any other form of life than this must be that spiritual condition which is not recognizable by our physical senses.

111

All forms of life are continually taking into themselves fresh increment of matter in some form. This fresh increment of matter is converted into other forms, for the support of the energies of that life which performed this conversion; and when this is accomplished, it must give way to fresh increments of matter, which undergo similar remoleculizations. The old must give place to the new. There is no such thing as standing still, except with certain definite provisions for temporary inactivity. Otherwise than this, inactivity is death. As new material is added, the old must be disposed of, must be cast out, or built up into formed material, where it takes no further part in the physiological activities of the organism. Continuous molecular change of matter is the law of physiological existence. In other words, the remoleculization of matter is necessary to the support of life; and when this remoleculization ceases to be performed by the life force, except under the provisions of rest, that life ceases to exist.

As the living form must be supported by fresh increment, so it must give back excrement. Life is not to be supported by the continuous remoleculization of the same matter. After it has once accomplished its function in the economy, it becomes unfitted for further physiological use, and is disposed of in one of two ways. It is either built into fixed material, or it is cast out as excrement. In either case it is disassociated with the physiological activity of the organism. The sum of the excrement, fixed and formed material, must, in all cases, equal the sum of the increment. There is nothing lost, nothing gained. All increment not present in the organism, must have been given back to the outside world as excrement, no matter what the form of life.

There is a general law to be observed in the formation of excrementitious matters. Liebig, while opposing the life-theory of the fermentations, gives expression to this, as a law of all fermentation. (I translate from Chemische Briefe, 6te Auflage S., 258. " Die Gährnng ist stets in ihrem Resultate eine Spaltung eines zusammen-gesetzten Atoms in eine sauerstoffreiche und eine sauerstoffarme Verbindung; indem sich in der Alkohol-gärung eine gewisse Quantität von sauerstoff von den elementen des zuckers in der form von Kohlensäure trent, erhalten wir den brennbaren, leicht entzündlich sauerstoffarmen Alkohol.") " Fermentation is always, in its results, a splitting of combined atoms into a compound rich in oxygen, and a compound poor in oxygen; so in alcoholic fermentation, a certain amount of oxygen is divided from the sugar in the form of carbonic acid, and we obtain the inflammable alcohol, poor in oxygen."

When we compare this statement with what is now known of fermentation we find it correct. But we may do more than this, for if we extend it so as to cover all life, in whatever form, we find that it is still correct; and that it is a law of the formation of waste products. For in the animal forms we find as excrementitious products the same carbonic acid rich in oxygen, and urea poor in oxygen. In the higher plants we still find the same carbonic acid rich in oxygen, and the alkaloids and organic acids poor in oxygen. Here we see as the final result of the remoleculizations by the life force, waste products analogous to those of the yeast plant. In other words we find in all forms of life a respiratory waste product rich in oxygen, and an urinary waste product poor in oxygen.

I have already shown that in the physiological sense, there is a close resemblance, a certain oneness of plan existing among all forms of life in the matter of taking food. Now it is my purpose to show that this identity of plan is extended to all the living forms in the matter of waste products as well.

While there are almost unending differences of outward conformation there is a continuous physiological sameness in them all. While certain cells are endowed with the power of combination for building wonderful forms, as in man or in the tiger, in the lofty palm or the forest oak, others, not possessing the power of organization into complex forms, fall apart and carry on their physiological processes singly, as in the yeast plant, or the mycoderma aceti. In the unicellular organisms, all the vital functions, so far as they are differentiated, are carried on in the single cell; and in the higher animals which proceed from the growth and development of some single, and equally minute germ, specialization of function goes hand in hand with specialization of physical form. Yet in all this specialization there is no radical change in the functions of the individual cells; nor is the sum total of the physiological phenomena modified in their nature. In the one matter of seeds, we are amazed and confused, by the never ending varieties of form; but in the examination of the physiology of seed forms and the awakening of the germ into active life, we find each to consist of a germ with the accompaniment of a store of formed food material for the nutrition of that germ, until it has formed the organs for the gathering of its own food; and the physiological means of using that store is practically the same in them all. If we examine the egg, the result is the same. Among the varieties of form there is but one plan.

In most cases, apparent new functions, seen as we ascend the scale of life, consist, when closely examined, of vicarious cell action. Not a new function, but a particular function performed for the community of cells, so to speak, by certain cells specialized for that purpose; as we have already seen in the matter of digestion. We find the specialization of muscular fibre for the performance of complex motion through which a portion of the cells of the body act for the whole, but it is not a new function, but, a function that has become specialized; for in the unicellular forms we have motion without muscular fibre.

So we might go on showing that the so-called specialized functions are more highly developed attributes that are to be found in unicellular life. But this is unnecessary.

Waste products may be divided into two classes: the respiratory and the urinary. The respiratory product is always rich in oxygen, while the urinary product is always comparatively poor in oxygen. This distinguishing feature remains constant for all forms of life, whether animal, vegetable, or the lower forms, that are so doubtful that some have thought it well to create for them a separate department.

The waste products of the animal forms are sufficiently well-known. The consideration of them has been entered into by every author who has written on the subject of physiology; and is therefore unnecessary here. I will, however, call attention to some important facts that will be of use in the consideration of the variations observed in the waste products of some of the lower forms of life. There is, perhaps, no part of the animal frame that is absolutely constant in its chemical components. All are subject to variations within certain limits that are by no means well defined. So it is with the waste products. While they are constant in their general characters, they are subject to variations with changes in the character of the food employed by the animal. For instance, if my urine is acid to-day, I may render it alkaline to-morrow, or next day, by eating a few oranges or lemons. The small amount of the alkaline base, combined with the acids of the fruit, is sufficient to bring about this change, while the acid itself is destroyed by remoleculization. This instance is doubly instructive, for if we

6*

take into the economy a simple substance, that substance, if not
thrown out of physiological activity, by being built into fixed
or formed material, must appear, in some form, in the waste
products; and in this instance it is found in the form of urate
of potassa and urate of soda; while the acid with which it was
associated, being dependent on its molecular structure for its
chemical characters, is destroyed by remoleculization.   Its
component elements appear, but in other molecular forms.
Hence, the effect on the waste product is not different from
that which would have been brought about if the alkaline
base alone had been taken.   This may be stated as a general
law for the organic acids combined with the alkaline bases.

All of the excrementitious products possess toxic, or poisonous properties in some degree.   If, by the occurrence of accident or disease, urea be retained in the blood of an animal, it
soon presents symptoms of toxæmia, and if the accumulated
urea is not speedily eliminated, the animal dies.   This is a
general law of the waste products.   No living being can sus-
tain life with a large percentage of its waste product retained
in its circulation.   Neither can a micro-organism continue to
grow after a certain amount of its waste products have accumulated in the menstruum in which it is placed.   *Urea may
be regarded as the alkaloid of the animal kingdom, and is the
analogue of the alkaloids of the vegetable kingdom.*

WASTE PRODUCTS OF PLANTS.

The consideration of the waste products of the vegetable
kingdom is especially difficult, for several reasons: Plants are
characterized by the large amount of their formed material,
as compared with the food material consumed by them in
their processes of vital activity; and therefore, the amount of
their waste products actually *excreted* is proportionately small.
In them, however, other means are found for freeing their

circulating fluids of waste materials, than that of actual excretion as seen in the animal forms, which will be considered
farther on.   There is, however, a sufficiently copious excretion
of the respiratory waste product, carbonic acid, in plants, and
in addition to this, oxygen.   These are too well known to
require further comment here.

There is, also, a true excretory function performed by the
roots of plants, and by the seeds in the process of germination.
"When barley, or other grain, is caused to germinate in pure
chalk, acetate of lime is uniformly found to be mixed with
it, after the germination is somewhat advanced. . . . .
In this case the acetic acid must have been given off (excreted)
by the young roots during the process of the germination of
the seed." (Vide Johnston's Agricultural Chemistry,
page 81.)

This well authenticated fact may be regarded as the foundation of the theory that plants are endowed with the power
of excretion.   It is supported by the authority of Decandolle,
and the convincing experiments of Macaire, although the experiments of others have shown that this excretion is limited
to very small amounts of matter.   Macaire seems to have
found opium in the soil in which the poppy plants grew. He
also found, in washing the soil with pure water, that it yielded
a considerable quantity of acetic acid and a trace of a brown
organic matter.

Liebig devotes considerable space to this, in his agricultural
chemistry, with especial reference to the possible effects of
these excrementitious substances on the rotation of crops.
Other chemists have also occupied themselves with it, and
while it has been definitely shown that in some special cases
plants may eliminate a sufficient quantity of excrementitious
matter to prove injurious to successive generations growing in
the same soil, the rule is that poisoning from this cause is not

to be expected. In many instances, however, it is found that individual plants are poisonous to others growing in their neighborhood. It is rare to see a very large black walnut tree that has not a clearing around it, wherever it may stand in the forest. This is not on account of its shade, but something eliminated by the tree that is hurtful to other trees. I remember well an effort to raise corn on the south side of a row of walnut trees. The experiment was continued for many years. The corn was injured seriously for many feet distant, where it was never shaded by the trees. Very many such instances are known. Enough has been demonstrated to show conclusively, that most of the substances formed in the fermentations are also excreted in very small amounts, by various plants. Even alcohol and ammonia are formed in this way, in small quantities. Although these substances, as a rule, contain no nitrogen, their chemical construction is such as to show their relationship to urea, and leave no doubt that they are properly analogues of that product in the animal kingdom.

ALKALOIDS.

By far the most important compounds to be considered in this connection are the alkaloids. These bear the same relation to plants that urea bears to the animal forms. There is seen in their chemical conformation and characteristic properties a marked relationship; and yet among them is found a wide divergence of poisonous effects. All possess, in some degree, the power of intoxication, while a few are the most virulent poisons known. Nearly all are composed, as urea, of carbon, hydrogen, nitrogen, and oxygen. In a few of them oxygen is lacking; and in all the nitrogen is in smaller proportion than in urea. As we descend the scale, the nitrogen disappears, the oxygen is increased, and we have the organic acids and alcohols. All of these are, properly, waste products of the vegetable kingdom, though they are not all excreted.

As we have already seen, the higher plants are marked by the apparently small amount of their waste products and the large amount of their formed material. The formed material presents two very marked characteristics, which will, perhaps be better understood if we designate them as formed material and "fixed" material. Then we may class the formed material as that which is elaborated and stored away for future use, such as the starch, oil, etc., surrounding the germ in the seeds, and in certain organs of the plant, and various other products designed for future employment by the vital energies of the organism. The "fixed" material, on the other hand, is such as is placed permanently beyond the vital energies of the organism. Some portions of the fixed material may still be of use to the plant by the physical support it gives to its organs; as lignin, which forms the stems and branches of trees, and the bark, which serves them as a protection. They are not, however, of any further physiological use to the plant after having once, or a few times, served as conduits for the circulating fluids.

Although waste products may be found in the fluids of the plant, as in the blood of the animal, it is in this fixed material that we find the bulk of these substances; and it is here that the toxic principles are found in greatest abundance. Instead of being excreted, thrown out to the outer world, they are stored in the disused cells of the wood and bark, united with other waste products in the form of insoluble compounds. This fact seems to have prevented an early recognition of their real nature.

Sachs, in his text-book of botany, describes them as "degradation products, which are no longer useful to the plant," and as "secondary products of metastasis." When we look over the products thus set apart by Sachs, we find that nearly all of them may be set down at once as analogues of the urine in the animals, of the alcohol, the acetic acid, the butyric acid, etc., of the fermentations.

It is to these products that we look for the most of the vegetable substances now employed as medicines. The active principles of these are the alkaloids. Therefore their effect upon the animal economy is well known to medical students, and need not be recited here. There are many other substances found in the same relationship to the vital energies of the plant, which are worthy of special consideration, if we had the time at our disposal. Tannic acid is one of the most constant of these products, and there are a large number of others not yet mentioned. We will only mention one other class of substances, the coloring matters which remain in the wood cells in a fixed condition, as a result of secondary metastasis. Precisely similar phenomena are seen in the Micrococcus Chromogenes, of which there are several varieties known. All such circumstances serve to point us forward in our efforts to explain the *mysteries* of nature.

BACTERIA.

When we come to consider the bacteria, and the allied forms of life, we are at once amazed and astounded at the wonderful power they possess in the *remoleculization* of matter. This is one of their especial characteristics. This seems to be the form of life in which the largest amount of food material is consumed, and the largest amount of waste products given back, with the least building of tissue. The largest amount of remoleculization with the least amount of formed or fixed material. A whole jar of milk is turned sour, every particle of its sugar of milk converted into lactic acid, and yet the amount of the formed material is so insignificant that it requires an acute and trained observer, provided with the best means of search, to find a trace of it. When a small amount of yeast is added to a solution of sugar, the whole of the sugar will be *remoleculized*, producing carbonic acid and alcohol within a few hours, with, apparently, no increase of the yeast. In this instance, however, the yeast has grown, and performed its function of remoleculization, under the most difficult circumstances, for the reason that it is deprived of the nitrogen necessary for the building up of its cellulose and albumen. Under these circumstances the amount of yeast may actually decrease, as was shown by Liebig many years ago. Pasteur seems to have proven that in this case the young yeast cells grow at the expense of the nitrogen in the old cells, and thus destroy them to such an extent that the whole quantity of the yeast is diminished, and the vital energies of the plant weakened. This example serves to show the wonderful power of these low organisms.

Then we may take the example of the destruction of the dead carcass. The history of the experimentation of the past shows conclusively that it does not decompose, if protected from these low organisms. Yet within a few days it is swarming with this form of life, in the presence of which it melts away like ice before a summer sun. It is first attacked by one form, which consumes and throws out its waste products as long as it can grow in the continually increasing quantity of its own excreta. Then it must give way to others that have already appeared on the scene of action, to which the waste products of the first are no hindrance. These, in turn, very soon give way to still other forms, which again remoleculize the waste products of all the former that may retain a semi-solid or a liquid form, with any residue of the original carcass that may still remain; and in an almost incredibly short time the whole carcass has disappeared. The labors of Pasteur have thrown much light upon this subject. The fact that the acetic acid plant will again remoleculize the waste product of the vinous yeast plant, alcohol, and after that the mucidines will still remoleculize the acetic acid, converting it, in turn, into other compounds, is full of instruction for all these processes. Why is it that the yeast plant cannot grow continuously in the same menstruum, if proper food material be furnished it? It is a well known fact that only a certain percentage of alcohol can be formed in this way, and that to obtain a greater percentage we must resort to distillation. The plant is choked by its own excreta, alcohol. If this could be eliminated the growth of the plant could be continued.

An example of this is found in the bacterium lactis. If lime is added to the solution, lactate of lime is formed and the excretory product of the plant, lactic acid, is artificially eliminated, and the plant is found to grow continuously in the same menstruum, as long as supplied with food material, and lime to fix or eliminate its excreta. The form of excretion employed by the higher plants is artificially produced. This law of the relation of excretory products to the life that formed them is universal. It applies with the same force to the micro-organisms that it does to the higher animals, or to man himself.

It would seem that the vegetable world is divided into two great classes of life forms, and that it is the office of the one to build up, and the office of the other to tear down. Of all the life forms, the higher plants have the greatest power of structure building, both as regards quantity, in comparison with the amount of the material consumed, and as to the actual quantity of structure formed; while, in the lower plants, the bacteria and their allied forms, exactly the opposite of this is found, in every particular. No other form of life consumes so much in comparison to the amount of structure built. They are emphatically the destroyers of organic forms; physiologically constructed to and for this end. Yet, they must digest their food; if it be solid, or semi-solid, it must be liquefied, in order to be absorbed. Then it must enter into the cells by the process of osmosis, and under the influence of the life force be converted into protoplasm—albuminoids—into the tissue of the growing plant.

After this come the waste products. How? The cell dies? No. This will not explain the phenomenon. If it were the death of the cells, we would find the chemical forms of the cells in the waste products. This is not the case. Death does not alter the chemical forms. Therefore, there is another remoleculization for the formation of the waste products; a vital process of shedding out the used material in two general forms. The life force letting go of the matter it has had in its grasp, and casting it out as no longer of use to it. The formation of waste products is not, in any true sense, a process of dying.

Life resides in matter which it has formed into such molecular groupings as will suit its purposes, but matter does not live. Electricity may be resident for a time in the Leyden jar, but such a jar is not electricity. It may be said to be electrified, and matter may be said to be vivified. If life discharges matter which it has had in its grasp, the act is a vital one, and as all changes in matter under the influence of the life force consist in remoleculization, so does this; hence, waste products are always in different molecular forms from the tissues from which they are cast off, though they consist of the same elements. It is by a vital process, then, that the waste products are separated from the living organism, and in the bacteria the general laws of physiology remain unchanged. Here, as elsewhere, we find the definite respiratory product, and the toxic or urinary product. A sufficient number of these have now been studied for us to gain some notion of them; yet our knowledge is still limited to a few forms among the bacteria and their allies. The best known is the yeast plant. In this we have alcohol, succinic acid and glycerine, as the urinary excrement. The bacterium lactis produces lactic acid; the mycoderma aceti produces acetic acid. These names properly represent so many families, each of which contains a number of varieties. Saccharine substances form suitable food for them all. Two of them will, however, grow equally well in substances not saccharine. Another form produces butyric acid; another, a kind of mucus; another, ammonia; while others still, produce as their waste products, all of the ill-smelling products of putrefaction. I append a table of those best known, in connection with a table of the waste products of other forms of life, for comparison.

TABULATED STATEMENT OF THE WASTE PRODUCTS OF VARIOUS ORGANISMS.

RESPIRATORY PRODUCTS.	TOXIC PRODUCTS.

ANIMALS.

Carbonic Acid......................CO_2	Urea.............................CH_4N_2O
	Creatine ....$C_4H_9N_3O_2$
	Creatinine...................$C_4H_7N_3O$
	Uric Acid...................$C_5H_4N_4O_3$
	Hippuric Acid.............C_9H_9N, O_3

PLANTS.

Carbonic Acid......................CO_2	Theine ....$C_8H_{10}N_4O_2$
Oxygen ....O	Veratrine ....$C_{32}H_{52}N_2O_8$
	Morphine ....$C_{17}H_{19}NO_3$
	Quinine...................$C_{20}H_{24}N_2O_2$
	Strychnine ....$C_{21}H_{22}N_2O_2$
	Atropine..................$C_{17}H_{23}NO_3$
	Piperine ....$C_{17}H_{19}NO_3$

RESPIRATORY PRODUCTS.

Carbonic Acid......................CO_2
Oxygen ....O

TOXIC PRODUCTS.

Nicotine......................$C_{10}H_{14}N_2$
Coniine........................$C_8H_{15}N$
Curarine......................$C_{10}H_{15}N$
Tannic Acid..............$C_{14}H_{10}O_9$
Acetic Acid.................$C_2H_4O_2$
Citric Acid.................$C_6H_8O_7$
Malic Acid.................$C_4H_6O_5$
Oxalic Acid................$C_2H_2O_4$

BACTERIA.

Alcoholic Fermentation.

Carbonic Acid...............CO_2

Alcohol........................... C_2H_6O
Succiuic Acid..............$C_4H_6O_4$
Glycerine ....$C_8H_8O_3$

Acetic Fermentation.

Carbonic Acid...............CO_2

Acetic Acid....................$C_2H_4O_2$

Lactic Fermentation.

Carbonic Acid...............CO_2

Lactic Acid....................$C_3H_6O_3$

Viscous Fermentation.

Carbonic Acid...............CO_2

Mannite ....$C_6H_{14}O_6$
Gum ....$C_{12}H_{20}O_{10}$

Tartaric Fermentation.

Carbonic Acid...............CO_2

Propionic Acid..............$C_8H_6O_2$
Hydrogen.............................H

Butyric Fermentation,

Carbonic Acid ....CO_2

Butyric Acid.................$C_4H_8O_2$
Hydrogen ....H

Ammoniacal Fermentation.

Carbonic Acid ....CO_2

Ammonia.......................NH_3

(These two products unite to form ammonia carbonate.)

Putrefactive Fermentation.

Carbonic Acid ....CO_2
(Complex.)

Valerianic Acid.............$C_5H_{10}O_2$
Ammonia ....NH_8
Sulphydric Acid...................HS
Fat Acids (various eqv.).....$C\ H\ O$

ALKALOIDS.

There is, perhaps, no other one set of organic compounds that have puzzled chemists more than those given in this table, or that have been and still continue to be of more interest to the scientific world. The table, as it stands, is only intended to give an outline view of these organic matters as a class, by setting before you fairly chosen specimens from the different forms of life. It is not intended to be in any wise exhaustive, but simply a comparison of these products. The similarity of their chemical construction will be apparent at a glance, and yet they differ greatly in their toxic properties. Some, as strychnine, are violent poisons to the animal kingdom. While others, as some of the acids, are mild irritants. Most of those from the vegetable world are useful as medicines, while many are highly prized as condiments. The members of the acid series vary from the alkaloids in that they contain no nitrogen, and they also vary as markedly in their toxic properties. Their close similarity to the alcohols will be noticed. It will also be noticed that some of the waste products of higher plants are the same as those of lower organisms. This, perhaps, would not be expected at first glance; yet when we come to consider that all are plants, we should expect a similarity. There is, perhaps, an important difference to be observed here. The alkaloids of the higher plants, from the nature of their mode of excretion, are thrown out of physiological activity, by being combined with the vegetable acids, mostly in the form of insoluble salts, and thus laid away in the "fixed material" of the plant. On the other hand, the animal alkaloids, urea, uric acid, etc., are eliminated in a soluble condition. In case of the micro-organisms, the waste products are very generally eliminated in a soluble condition also. There seem to be some exceptions to this rule, as in the chromogenes. To what extent this may occur, does not yet appear.

But the comparatively more abundant waste product of the lower forms is of much importance in this connection. Those given are from the few that have been best studied. In these no proper alkaloids have, as yet, been made out by isolation and exact analysis, but we cannot argue, from this circumstance, that none of the lower organisms produce true alkaloids. As yet, but very few of them have been accurately studied, and these, for the most part, have been those that have been found useful rather than poisonous. Some examinations already made show a very near approach to the determination of the nature of some of these poisonous substances; and although no sufficient chemical analysis has yet been made, the action of the substances upon the animal economy, so far as trial has been had, go to show the similarity of these poisons to the alkaloids from the higher plants. This has been remarked by many experimenters, as I shall show presently. Therefore, we have much reason to believe that true alkaloids, of a poisonous nature, will be isolated at no distant day. We have already abundant evidence that poisonous properties are developed by a number of these forms. Otherwise, how can we account for the results of the Bacillus Anthracis? How else can we account for sepsis, on the germ theory, except to suppose that a poisonous alkaloid is one of the products of the organism? How else can we explain Dr. Koch's results in the production of gangrene in mice, but to suppose that some product of the remoleculizations by the organisms spread among the tissues of the animals and destroyed them wherever they went. He found that after they had once made a beginning, the tissues were always destroyed in advance of the growing organisms, without contact. This strongly illustrates the idea of the formation of a poison which is thrown out from the organism. Again, in the matter of sepsis, already referred to, it has been abundantly shown that the fluids contaminated by the micrococci are poisonous, when separated from the organisms; thus proving, again, that there is an organic poison developed, which is fully capable of producing its effects in the absence of the life that produced it, the same as is seen in the alkaloids from the higher plants. This has taken the name of sepsin, after the order of the naming of the vegetable poisons; although it has not, as yet, been properly isolated and examined.

SEVENTH LECTURE.

POISONS.

In this lecture I propose to examine more particularly the evidence given by experimenters of the finding of the poisonous products of micro-organisms.

Koch, after injecting putrid blood into mice and noting that a large amount killed by direct poisoning, while a very small amount produced disease, says of the first: "No inflammation can be observed in the neighborhood of the place of injection. The internal organs are also unaltered. If the blood taken from the right auricle be introduced into another mouse, no effect is produced. Bacteria cannot be found in any of the internal organs, nor in the blood of the heart. An infectious disease has not been induced as the result of the injection. On the other hand, there can be no doubt that the death of the animal was due to the soluble poison, sepsin, which has been shown by the researches of Bergman, Panum, and various other investigators, to exist in putrid blood. The animal has accordingly died, not from an infectious disease, but from a chemical poison. This supposition is confirmed by the fact that when less fluid is introduced into the animal the symptoms of poisoning are less marked, and are quite absent when one, or at most, two drops are injected. After the use of such a small quantity of blood, mice often remain permanently without any morbid symptoms. But a third of them will become ill after the lapse of about twenty-four hours, during which time they have remained perfectly healthy. The symptoms which are then present are characteristic and constant, and are in no case preceded by any of the symptoms of poisoning previously described."

These experiments show very distinctly that the experimenter was dealing with a diffusible poison of virulent character, produced by the organisms, but acting without them, as the alkaloids from the higher plants would act.

Dr. Sternberg says (page 257): "It is not alone by invading the blood or tissues that bacteria exhibit pathogenic powers. Chemical products evolved during their vital activity, external to the body, or in abscesses and in suppurating wounds, or in the alimentary canal, may doubtless be absorbed, and exercise an injurious effect upon the animal economy. Indeed, we have experimental evidence that most potent poisons are produced during the putrefactive decomposition of organic matter. The poisons *resembling the vegetable alkaloids in their reactions* called ptomains by Selmi, who first obtained them from the cadaver, are fatal to animals in extremely minute doses."

Klebs said, at the International Medical Congress at London, that the effects of micro-organisms were probably due to fine chemical workings. Very many of the experimenters in this field have spoken in a like manner of their findings, all of which go to show that the worker in this line of research soon comes to feel that he is dealing with a poison evolved by the organisms he has under observation.

Gradle says (page 66): "The successive chemical stages of the putrefactive change have as yet been incompletely traced. The changes which albumen undergoes resemble at first the process of digestion. It is converted into soluble forms, partly peptone, and then split up into leucin and tyrosin. Subsequently numerous volatile fatty acids, and various polyatomic alcohols (phenol, skatol, indol), appear, as well as a host of other substances in traces. Amongst them are a variety

of poisonous agents, especially some alkaloids (ptomains)."
The most constant bacterial form met with in putrefaction is
the bacterium termo. Some attempts have been made to isolate and examine septic poisons. This, so far, has only served
to show more positively the production of poisons and the
nature of their action. Panum found the poison formed in
the decomposition of nitrogenous substances was reduced in
virulence about one-fifth by eleven hours continuous boiling
of the infusion. This continuous boiling was certainly sufficient to show that the poisonous effects were not due to living
micro-organisms remaining in the fluid. In this case, moreover, some flakes of coagulated albumen were found in the
liquid, which upon trial proved to contain a very virulent
poison. This fully accounts for the loss of virulence by the
liquid. This experimentation shows that the substance had
none of the characters of a molecular motion poison described
by Baron Liebig; nor is it a decomposing substance or a
ferment, as each and all of these are destroyed by heat. The
chemical structure of the substance is of a more stable nature.

There is a sharp distinction manifest between the action of
these poisons when developed in the animal body by the remoleculization of the organisms in situ, and when developed
out of the body, collected and administered separate from the
organisms. In the latter case, the effects produced are similar
in their general character to that of the alkaloids from the
higher plants, that is, comparatively speaking, they are immediate. The disturbance is usually manifest within a few
hours at most; and if the animal withstands this immediate
effect of the poison it returns to health, i. e., all the symptoms
pass away and no disease is produced ; whereas, if the micro-
organisms which elaborate the poison are planted in the
tissues of the animal and grow, there is no such immediate
poisoning. On the contrary, it remains well for a day or

7

more, and then the symptoms come on more slowly and gradually increase in violence usually until death is induced. A progressive disease is the result in the one case, an immediate, but transient poisoning in the other. This difference in the results of inoculations made with fluids rendered poisonous by the development of micro-organisms within them, and similar inoculations with fluids containing micro-organisms but without sufficient developed poison to produce immediate effects, has been marked by a large number of investigators and is now perfectly well known.

Putrefaction does not always result in the formation of virulent poisons. Many times meats, or meat infusions, may rot away, and poisons of a very marked character cannot be found at any stage of the process. This difference must be explained by the fact that the same micro-organisms have not grown in the one as in the other. If the ordinary bacteria of decomposition are accompanied by a form that produces a virulent alkaloid as a result of its remoleculizations, the poison will be found, otherwise, it will not. Just as in a mass of plants, one, or a few only of the many, will produce a virulent poison, though all grow in the same soil. Again, as we have said, the mycoderma aceti will feed upon the alcohol formed as the waste product of the vinous yeast plant. In the same way other micro-organisms may feed upon and destroy, through their remoleculizations, the poisonous products of those that may have gone before them. Hence, it is found that septic cadavers are apt to lose their septic characters as decomposition advances. These facts, while they serve to illustrate the manner of the formation of poisons, also illustrate the extreme complexity of the subject.

FARTHER INVESTIGATION NEEDED.

Many other instances of a like nature might be brought forward, if time would permit, but it is no part of my intention to treat this subject exhaustively. Some of the cases seem to show that the poison is of the nature of a local irritant only, as appears in Dr. Koch's demonstration of the bacillus tuberculosis; while many others seem to be general poisons, with special tendencies to particular organs, as is seen in the splenic fever produced by the bacillus anthracis. This is sufficient to show that there is much yet to be learned in this direction; indeed, that this work is as yet in its infancy. In truth, the work done thus far has been in the nature of proving that organisms are regularly connected with the fermentations and decompositions, and also with certain processes of disease, rather than any effort to show how they bring about these results. This was necessary in the first instance, as the forerunner of a farther and more complete understanding of the subject that must come from a closer study of the modes through which these low organisms produce their effects. It is plain that we cannot know any one of the disease germs until all of the products of its remoleculizations have been isolated and studied separate and apart from the organism itself, as has been done with the products of so many of the higher animals and plants. I see no other plan by which we can ever know the capabilities and all of the possibilities of micro-organisms, in connection with the production of disease. In most cases of the discovery of poisonous properties in the higher plants, the specific product of the plant in which that poison resides has proven itself capable of isolation by the means known to chemists; and the disease-producing poisons of the lower organisms will do the same when the effort to find them has been prosecuted with sufficient skill and energy. As alcohol has been isolated, as the vegetable alkaloids have

been isolated, as the toxic elements of the waste products of
animal life have been isolated, and as, through this isolation,
the exact properties of these agents have become known to
us, so must sepsin be isolated, so must the poisonous product
of the bacillus anthracis (Anthracine) be isolated, so; must
the poisonous principle of every disease-producing germ be
isolated, and each of these studied separate and apart from
the organisms which produce them; and in each case their
properties must be demonstrated and determined by direct
experiment, before they can be said to have been accurately
studied, and that we know their properties and capabilities.
This is the recognized means of learning the powers and
capabilities of the products of the higher plants and animals.
No one is foolish enough to deny the value, the necessity, of
the knowledge to be gained by this mode of study in the
vegetable kingdom.   If there has been any truth arrived at
in all the research that has been given to the subject of the
production of disease by the lower organisms, that truth
requires to be extended, and rendered more exact, by the
modes of study we have just indicated.

PHYSIOLOGICAL CHANGES.

The question, as to whether the nature of the bacteria and
their allied forms are always essentially the same, is one that is
very important to determine. Very many persons seem to think
that the characteristics of this or that *organism may be permanently changed*, by temporary changes in the *media* in which it is
grown. This is a question of the gravest importance; and one
on which there has, as yet, been very little or no exact experiment.   Therefore we have nothing but conjecture; unless,
indeed, we may regard the experiments of Pasteur, in his
efforts to moderate the virulence of the bacillus anthracis, the
organism of chicken cholera, etc., experiments in this direction.

These, however, certainly cannot be classed as exact experiments. Yet so far as they have scientific value, it is undoubtedly in the direction of the maintenance of the proposition that the nature of organisms may be profoundly changed by temporarily changing the media in which they are grown; for it is on this basis that we must account for changes in results. Such a proposition will, however, require the most rigid proof before we can accept it as a modification of the essential nature of the products, that will be carried, with the organism, back into the media from which it was transplanted.

The other proposition, that the products of the organism *change with the media* in which it is grown, is of a very different nature, and much more plausible. We all know that our own urine may be changed from acid to alkaline, by certain changes in diet; that the amount, and to some extent, the quality of opium, is affected by the nature of the soil in which the plants are grown. Still it does not appear in these cases that the essential features of the waste products are more than modified in some unimportant particulars. The urine will still furnish urea, and the opium will still furnish morphine, although each will, possibly, furnish their characteristic product in diminished quantity. But in either case, the original character of the product will be resumed with the resumption of the original diet.

One would scarcely expect the yeast plant to produce a nitrogenous alkaloid when grown in a pure solution of sugar or starch, containing none of the element nitrogen. So an organism whose normal habitat is nitrogenous compounds, and whose waste products are nitrogenous, if transferred and found to grow in pure starch, would certainly not produce a nitrogenous alkaloid as a part of its waste product, while growing in the starch; yet I would expect it to again produce the nitrogenous product if replaced into nitrogenous matter;

if, indeed, the plant retained sufficient vitality to perform its functions. The material for the formation of the waste product must be present in the food, otherwise it cannot be formed. Again, it may be that the remoleculization will be different in foods of a different nature, even though the elements be all present. Dr. Miller informs us that, in his experiments, he has found that the bacterium lactis produced no acid when grown in soup; but when a little sugar or starch was added, the lactic acid was promptly formed. He gives us no intimation of the nature of the waste product of this organism when grown in the soup not sweetened. This however, serves to show how the waste product may be changed, temporarily, by changes in the nature of the food. That changes of this nature may occur in a large number of the low forms of life seems not only plausible, but probable. But that this will affect the nature of the organism permanently, is another question; and one that will require much experiment to prove satisfactorily.

MANNER OF ACTION.

How can these low organisms produce disease? This question is being asked by many of the thinking men of the world to-day. Men seem to be at a loss for a reasonable answer; such an answer as will satisfy the mind that is earnestly seeking an explanation of the phenomena described by such experimentalists as Koch and Pasteur. These men have not answered it. They tell us what they have found; that certain phenomena invariably follow the introduction of certain germs into the bodies of animals; that certain germs produce certain diseases. How do they accomplish this? what is their modus operandi? This question must have a reasonable answer before the germ theory of disease can find a firm basis in the minds of the masses of men. The days of hocus-pocus have passed away from the medical profession forever. A few men may take these things on faith, and wait for time to develop the rational answer, but the many will wait, and watch for farther developments to point out a reasonable and philosophic answer, before they will accept this theory without reserve. Can such an answer be given with our present knowledge of the subject? If it be required that every step be proven by actual demonstrated fact; if reasonable circumstantial evidence be not admitted, only a few cases can be clearly made out. But on the other hand, if the physiology, as previously explained, be accepted, the manner of the production of disease becomes clear. I should not insist that the poisonous substance must, in every case, be an alkaloid. But it must be an organic compound closely akin to an alkaloid. To say that these organisms produce disease, simply by their presence, is not a sufficient explanation. We all know that their presence is not, in itself, a sufficient cause; for if this were so, one organism would be as potent as another. And we all know, who have had any experience in the examination of these forms, that many of them are without any evil effect whatever; that wounds in the mouth and elsewhere heal readily and perfectly when covered with them. We also know that the great majority of the higher plants are innocent of any evil effect upon the human system; and it is perfectly reasonable that we should find the same differences to exist among the lower plants. Many kinds of plants grow together without injury to each other; but plants are found that destroy their neighbors by the poisonous effects of their waste products. Then it seems plain that the differences must be found in the peculiar products to which the particular organism gives rise. These we have already sufficiently explained. We have, also, sufficiently explained the modus operandi, in the destruction of organic substances in general. They are simply remoleculized, as any other food material is remoleculized; nothing more and nothing less; and it is in the chemical re-organization of the elements for the formation of the waste products that poisons are evolved.

It seems perfectly natural that in this remoleculization there should often be a residue that escapes digestion; as is seen in the so-called rotting of wood and many other substances. It is not probable that one organism is capable of the complete destruction of such substances; but rather, that many are concerned in the work.

I may be permitted to make a suggestion in regard to the cause of the difference, that is so often noticed, of the comparative liability of different persons to attacks by these organisms. Several persons receive wounds. So far as can be seen, their chances for a speedy recovery are equally good; but some of these develop sepsis, while others do not. Now, much of this is explained by the fact that we are unable to perceive the entrance of these organisms, and they may gain admission to the one and not to the other, and we be none the wiser until we find that sepsis is established. This is always a possible explanation, for the history of the experimentation on this subject shows full well that we have not, even yet, learned to prevent the ingress of these germs with absolute certainty. Yet over and above all of this, there is a certain residue of cases that show that different persons are liable in different degrees to sepsis. This has been noticed ever since there was such a profession as medicine, and the class of patients to which it refers have been pretty closely defined. They are those whose powers of life have been weakened, by whatever cause. This is said to be the resistance of the tissues to the invasion of the organisms. How do they resist? Now, if we accept the explanation of the formation of the digestive bodies, as previously explained, this supposition becomes tenable. When an organism or spore of an organism falls upon a cell or naked granulation of a wound, and finds there the conditions suitable for its development, its digestive body is at once thrown out for the preparation of its food material which exists in the tissues with which it is surrounded. This digestive substance proves irritant to the cell upon which the germ has fallen. This irritation excites the cell to throw out its peculiar digestive body for the removal of the irritant with which it is beset. Here we get a glimpse of the antagonists. It is cell against cell; digestive body against digestive body. A contest has begun. Which will win? The stronger, of course. If the man be vigorous and healthy and his tissues highly endowed with the energies of life, the chances are this much in his favor; but if, on the other hand, his powers of life be at a low ebb, if his tissues respond to irritants but feebly, then the chances are, this far, in the favor of the intruding organism. In the first case, the cell upon which the intruding germ has fallen responds promptly to the irritation; a substance calculated to free it from the effects of the intruder is formed and poured out upon it and meets the digestive body of the intruder, hindering, dissipating and nullifying its action. In this way the intruding germ is, in many cases, possibly, overcome and driven out. In the opposite case, where the vital powers are at a low ebb, the response to the irritation is, perhaps, very sluggish, the digestive body is poor in quality and scanty in quantity, and the intruding germ has an easy victory.

If we suppose that the intruder is a spore, a seed, with a store of food material laid up within its shell, the case is not different essentially. The history of the experiment on this subject shows that during this process excretory products are thrown off, and these may prove direct irritants that will arouse the resistant energies of the tissues. Surely, if the tissues are capable of forming, by reason of irritation, a secretion that will digest a piece of ivory that has been thrust into the flesh, which has been proven by direct experiment, we should expect this kind of resistance to be offered to the development of disease-producing germs.

Facts of a very decisive nature bearing on this point have been brought to light by Dr. Sternberg, in the following experimentation, which I quote :—

" If we add a small quantity of a culture-fluid containing the bacteria of putrefaction to the blood of an animal, withdrawn from the circulation into a proper receptacle, and maintained in a culture-oven at blood heat, we will find that these bacteria multiply abundantly, and evidence of putrefactive decomposition will soon be perceived. But if we inject a like quantity of the culture-fluid with its containing bacteria into the circulation of a living animal, not only does no increase and no putrefactive change occur, but the bacteria introduced quickly disappear, and at the end of an hour or two the most careful microscopical observation will not reveal the presence of a single bacterium. This difference we ascribe to the vital properties of the fluid as contained in the vessels of a living animal; and it seems probable that the little masses of protoplasm, known as white-blood corpuscles, are the essential histological elements of the fluid, so far as any manifestation of vitality is concerned.

The writer has elsewhere suggested that the disappearance of the bacteria from the circulation, in the experiment above referred to, may be effected by the white corpuscles, which, it is well-known, pick up, after the manner of amœbæ, any particles, organic or inorganic, which come in their way. And it requires no great stretch of credulity to believe that they may, like an amœba, digest and assimilate the protoplasm of the captured bacterium, thus putting an end to the possibility of its doing any harm.

In the case of a pathogenic organism we may imagine that, when captured in this way, it may share a like fate, if the captor is not *paralyzed by some potent poison evolved from it* (the italics are mine), or overwhelmed by its superior vigor and rapid multiplication. In the latter event, the active career of our conservative white corpuscle would be quickly terminated, and its protoplasm would serve as food for the enemy. It is evident that in a contest of this kind the balance of power would depend upon circumstances relating to the inherited vital characteristics of the invading parasite and of the invaded leucocyte."

In these paragraphs we have the strongest evidence of the truth of the supposition stated, of the nature of the combat between the tissues or the white corpuscles of the blood and the wandering cells for the tissues, on the one hand, and the invading organisms on the other. The writer, though not intending to put forward his own experimentation in this work, may say that he has also seen strong evidence of the truth of this in tissue taken directly from man to the warmed stage of the microscope, in which the wandering cells were found loaded with micrococci, which in many instances seemed to be destroying the cells. Some were motionless and filled to overflowing with the organisms, with little chains of the micrococci extending from them, while others containing but few of the organisms exhibited their usual motions. This phenomenon may occasionally be demonstrated in the peculiar granulations which are sometimes found under plates for artificial teeth, where the gums have taken on a bad condition.

Dr. Koch has also seen micrococci in the white blood corpuscles, under circumstances that indicated that the blood cells were being destroyed by them. He says: "Their relation to the white blood corpuscles is very peculiar. They penetrate into these and multiply in their interior. One often finds that there is hardly a single white corpuscle in the interior of which bacilli cannot be seen. Many contain isolated bacilli only; others have thick masses in their interior, the nucleus being still recognizable; while in others the nucleus can no longer be distinguished; and finally, the corpuscle may become a cluster of bacilli, breaking up at the margin—the origin of which one could not have explained had there been no opportunity of seeing all the intermediate steps between the intact white corpuscles and these masses."

With these facts before us, and with the plain teachings of the relations of the life force to matter, as exemplified in the phenomena of digestion, nutrition and the formation of waste products, it seems to me we must not fail to gain an understanding of these processes. The cells attacked, either digest the invading germs, or the invading germs digest them. And here, as everywhere else in nature, the stronger will be the victor. In other cases, as in septicæmia, the excretions of the bacilli do not seem to produce such marked local lesions, but the waste products of the organisms are absorbed and act the part of a diffusible poison, producing the general symptoms manifested in these cases.

Again, the history of cases shows that sepsis is most liable to occur in the early time of the healing of wounds; at a time when the tissues may be supposed to be still suffering from the effects of the shock. In this case it must be supposed that the tissues are not so well able to contend with the intruding germs as they are afterward, when the process of granulation is going forward vigorously.

We have also experimental evidence establishing this point. Sonnenschein found that certain bacteria with which he was experimenting failed to grow in the tissues of an animal. Believing, from the results of previous experimentation, that sepsin injected with the bacteria favored their development, and having found the effects of sepsin similar to that of sulphate of atropia, he injected a small amount of that drug with the organisms and found that, under these conditions, they grew and produced sepsis. It is difficult to see how the sulphate of atropia could aid the bacteria otherwise than by paralyzing the tissues temporarily, preventing their resistance until the organisms had established themselves. A number of other experiments have developed facts of a similar nature.

Gradle says, page 123, "We do not know whether putrefactive bacteria exert any direct influence upon the exposed tissues. Even if that be not the case, the products of decomposition which they engender irritate the wound in an unmistakable manner. This is seen in the redness and sensitiveness of the margins of the wounds. . . . . These bacteria are always overcome by the tissues."

INFLAMMATIONS.

The question as to whether we ever have the formation of pus without the presence of micro-organisms, has been much discussed, pro and con, of late years, by men standing high in the medical profession. It has even been contended that we never have inflammation without them, notwithstanding the well-known effects of the blister and the mustard plaster. Did it ever occur to any, that in the blister and the mustard plaster, we are making use of an irritant that has been prepared for us by the life force, and is, to say the least, closely akin to the irritant that we would expect from the disease-producing organisms? But have we not mineral irritants, and may we not produce inflammation by their use? Is not pus a result of inflammation? I am not unmindful of the fact that there are several theories explaining the production of pus; and that pus does occur under some conditions in which the presence of inflammation, at any time previous to its formation, is extremely difficult of demonstration. Yet I am of the opinion that there are very few surgeons to-day who do not regard the formation of pus as one of the results of inflammation, and who will not naturally expect the formation of pus to result in case of prolonged and intense inflammatory action.

Still, the formation of pus does not necessarily occur as a result of every severe inflammatory state; and some recent experiments by M. I. Straus and others, in both France and Germany, seem to show that pus will not form in inflammation artificially excited, if micro-organisms be excluded. Other experimenters have had other results, and the question is not yet settled by the experimental method; and we might with some reason say that it cannot be so settled, until the methods of experimentation approach more nearly perfection than they seem to be at the present time.

However, just so long as we regard the formation of pus to be a result of inflammation, must we expect that pus will be formed without the presence of micro-organisms. If this is not the true explanation of this phenomenon, if future developments should demonstrate that these two processes are not related to each other as cause and effect, then it may, possibly, be shown that pus cannot form without the presence of micro-organisms.

At the same time it must be admitted that the formation of pus is generally accompanied by the development of micro-organisms, and that their presence facilitates pus formation. This is clearly the result, not of temporary, or a single application of an irritant, but of the continuous irritation by the secretion or excretion continuously given out by the organisms; and we have no reason, that I know of, to suppose that the result would be much different if the same products of these life forms could be gathered and continuously distributed in the tissues artificially.

In many of Straus' experiments, and others of this class referred to above, agents were introduced that evidently paralyzed the tissues, and as pus formation is a vital act, it was necessarily prevented. Every such element must be rigorously excluded from our experimentation before the results can be depended upon as conclusive.

BLOOD DISEASES.

The production of disease, however, is not limited to inflammation and pus formation. Many of the diseases now supposed to be parasitic in their causation are not essentially inflammatory, but are known as systemic or blood diseases. This class of diseases are not caused by any simple irritant, but by some toxic element developed by the organisms. As to what this toxic element may be we have only conjecture, but after what we have just said in reviewing the waste products of the various life forms, this conjecture is a close approach to knowledge. The elements of the compounds are pretty certain to be written C H N O, in varying equivalents. It is possible that the N may be left out of some, or that the O may be left out of others. But this is not very probable. We know enough of the alkaloids, to know that they may become a potent cause of disease, if developed continuously in the blood or tissues.

It may be said that we know of no alkaloids that will produce the effects shown in the diseases that we see around us. This is very true. A few years ago we knew of no alkaloid that would produce the effect of woorara, or quinine, or atropine, or strychnia; and when we have discovered others, and have learned their effects, they will be as easily understood as those now known to us. That systemic disease may result in this way, is not only possible, but it

seems very probable that this will be the final solution of
the problem.   Indeed, we may say that from all that is now
known, this seems certain.   But it is not from the formation
of the alkaloids alone that we will find these effects produced.
The organic acids will be found to have their share in the
work; also various irritants, and the digestive bodies developed by the organisms.   It will be found that all of these
products are formed together and do their work together,
producing the complex results that we see in the complicated
forms of disease so frequently met with.

While the mere physical presence of organisms will in no
case account for the effects ascribed to them in the production
of disease, they cannot be entirely ignored.   When bacteria
or micrococci are gathered together in groups and large
masses (zoöglia) in the tissues, they will certainly have some
evil influence by their mere presence in such situations,
especially when this happens to be in important organs.
Also, the fixed material laid up in these cells may, after the
death of the organisms, have its influence.   But I am persuaded that neither of these causes is the potent factor in the
production of disease.

I would not have you suppose, however, that we can have
no disease without micro-organisms.   Man may be abused by
the life forms around him, and man may abuse himself.   He
may eat too much good food and suffer in consequence.   He
may injudiciously expose himself to the inclemencies of the
weather, and start serious inflammation of important organs.
He may exhaust his energies by overwork, and suffer all the
consequences that anæmia brings in its terrible train.   He
may suffer from faults in the physiological activity of his
own tissues, from neoplasms, as cancer sarcoma, lipoma, and
very many other affections that the micro-organisms do not
have to answer for.

APPENDIX.

DENTAL CARIES.

ITS RELATIONS TO THE GERM THEORY OF DISEASE.

Many theories have been advanced, in times past, to account for caries of the teeth. Most of these have been vague and indefinite, and have passed away with the advance of time. About the beginning of the present century the vital theory, as it was called, was prominent. By those who held this theory, caries was supposed to result from an inflammation of the structure of the dentine, which terminated in the final breaking down of the part; and as this structure is incapable of physiological repair, a cavity was the inevitable result. This theory seems to have been very thoroughly disproved by the following considerations. It was observed that artificial teeth, constructed of ivory, bone, sheep's teeth and human teeth, were as liable to caries as the natural teeth; and that this decay was, to all appearance, the result of similar, if not precisely the same causes. This, of course, could not be the result of vital forces existing in the structure undergoing the process of decay. The conditions of the decaying portions were very closely studied, to see whether or not anything could be discovered that would show that these processes were essentially different from each other. Some differences were found, but the more they were studied the more evident did it seem that these differences were not of such a nature as to show that the processes were in any wise distinct from each other. These studies caused the abandonment of the vital theory; for, if dead substances decayed the same as living ones, the forces which bring about the result must be other

149

than the vitality of the suffering organ, and therefore cannot be the result of inflammation; it must be some force external to the tooth, something which attacks the tooth from without. This point seems to have been very thoroughly established, and we supposed it was well settled in the minds of all modern thinking men. But, curiously enough, the old notion has recently been revived by certain gentlemen in New York. The effort to maintain such an hypothesis, however, must inevitably result in failure; for in it there is a plain want of consideration of the known facts of the subject. It need not, therefore, detain us longer.

Caries of the teeth has been defined as a molecular disintegration of the tooth's substance, or a breaking down of the chemical constituents of the tooth, molecule by molecule. This destruction always has its beginning on the surface of the tooth, or in some pit, crevice, or other imperfection in the enamel. And it spreads from this point, as the focus, in every direction, the dentine being destroyed more rapidly than the enamel; hence, it usually happens that the cavity is larger within than on the surface of the tooth. Caries does not seem to be a simple solution of the tooth's substance; sometimes we find nearly all of the material removed from the cavity, in other cases we find the dentine reduced to a pulpy or semi-gelatinous mass, in which the structure of the dentine is more or less perfectly preserved. Some decays are white, some are black, some have a yellow tinge—all the shades from white to black may be found. But it was not my intention to give a lengthy description of the results of caries, but rather to confine myself as closely as possible to the discussion of the probable cause of the affection; I say probable cause, for I do not assume that the cause or causes are certainly known. Indeed, we may say that at the present time there is the greatest disagreement among even the best

informed men on this important subject; and that it is still
very uncertain that any of the theories in regard to the matter
are correct. It is proper that I should say, however, that
more than one of these explain the phenomena with sufficient
accuracy to be of great value, both in the prevention and
treatment of this affection. It must not be supposed that a
theory must be absolutely correct to be of use. Theories are
usually contrived in the effort to explain phenomena, and it
often happens that a false theory leads to as good an application of means to ends as the true one would do. Of this,
however, we can never have any assurance; therefore, as long
as there is a reasonable doubt the search for the truth should
continue.

The theory for the explanation of caries which has received
the greatest attention and the widest recognition in modern
times, is what is known as the acid theory. This theory seems
to account for the phenomena more perfectly than any other
that has as yet attained prominence in the minds of thinking
men. As a working theory, a basis upon which to found
principles of treatment, it has undoubtedly been the means of
good. Yet, in the scientific aspect of the subject, there is
much objection to be urged against it. A very large amount
of work has been done with the view of demonstrating the
absolute truth of this theory, all of which must be regarded
as a failure, so far as the attainment of that particular object
is concerned. The labor has not been lost, however; but, on
the other hand, has been of immense value. It is this labor,
the basis of fact which it has brought to light, that will be of
most service to us in the building up of other theories for the
explanation of the phenomena, which may serve us usefully,
until such time as theories shall be displaced by demonstration, the goal to which we are all looking forward.

According to the chemical theory, the substance of the

tooth is decomposed by an acid ; this acid acts more readily on dentine than upon the enamel, therefore the tendency to the enlargement of the cavity toward the internal portions of the tooth. Some writers, as Dr. Watt, have attempted to define the acids thus acting, and to divide decays into classes, according as this or that acid is active in its production. The acids that have been thus pointed out are : Nitric acid (white decay), Sulphuric acid (black decay) and Chlorohydric acid (intermediate colors).

Most of those who have written on this subject, however, have been content without specifying the particular acids so definitely; while some have been of the opinion that these particular acids have little or nothing to do with the matter, and seek to show that other acids are more likely to do the ugly work.

The origin of the acids which produce caries has been a subject of much inquiry. A great amount of labor has been bestowed upon this point. It has been supposed that the saliva contained it, and very careful examinations have been made in this direction, which have contributed to our knowledge of this secretion, but have thrown very little light on the point in issue, other than to demonstrate that the cause of decay is not to be found in this direction. It has shown that decay occurs in mouths in which the saliva is habitually normal; and that decay does not occur in some mouths in which the saliva is habitually acid.

The hypothesis that the acid is furnished upon the spot, through the decomposition of the food lodging between the teeth, or in crevices and imperfections in the enamel, seems much more feasible, and gives a much more satisfactory explanation of the phenomena. According to this hypothesis, the acid is formed in juxtaposition with the portion of the tooth suffering from its effects, even within the cavity itself, while the general fluids of the mouth may be neutral, or even alkaline.

The manner of the production of this acid has been a subject of much inquiry. All, I believe, admit that it must come in some way through fermentation or decomposition; at any rate, through a remoleculization of the substances lodged about the teeth. This leads directly to the discussion of the much vexed question of the fermentations and decompositions, and the relations of the life force thereto. The research of the last few years leaves but little doubt on this point. These processes are the result of the activity of living organisms. The teachings of Liebig on this subject may now be regarded as disproved. With this point settled, the old acid theory glides easily into the new germ theory, and we may recommence our studies from the new standpoint we have gained, carrying with us all the facts that have been developed to guide us on our way.

In the study of this subject many new questions spring to the front, and demand a hearing. Most of these have been discussed in a former lecture; but it now remains for us to study their application to this particular subject. Before doing this, however, it will be well for us to make a hasty review of the discussions that have been had in comparatively recent times on this point.

THE GERM THEORY OF CARIES.

Who may have first suggested that caries of the teeth is caused by living germs, I have no definite knowledge. The first work of importance upon this subject was by Lieber and Rottenstein, in Germany, which appeared in 1868. These gentlemen, in the prosecution of their studies of this question came to the conclusion that caries must result from the activity of micro-organisms through the production of acids.

This acid was, of course, produced in the process of fermentation; but in the light of the developments of the last few years, their observations seem vague and very indefinite. They attributed this result to leptothrix buccalis. The observations upon which their views were based were given at length; so that we are able to follow them through all their operations, and from the care with which these were conducted, judge of the merits of their deductions from them. When we pass their work in review carefully, we must admit that it is wanting in that extreme care that is necessary in investigations of this delicate nature; especially is this the case when we examine the work in the light that succeeding years have thrown upon investigations of this class. In fact, the modes employed in such work have been completely revolutionized since this work appeared; and as the general nature of such work has been discussed in a previous lecture, and its progress described, it need not detain us now.

Notwithstanding the evident errors of the work of Lieber and Rottenstein, it undoubtedly made a deep impression on the thought of the dental profession; and this question has not been lost sight of for an instant. While the opinions given were not accepted, and were, apparently, quickly disproved, it lent an impetus to the study; and though no great work has appeared, it has been progressing in an intermittent way, until now the profession seem to be ready to receive any work on this subject that will bear reasonably rigid criticism. Such a work is, fortunately, now appearing, from the pen of Dr. Miller, who, though an American, is working in the laboratory of Dr. Koch of Berlin. Dr. Miller is, fortunately, in the very best position possible for the performance of this difficult form of investigation; and the papers that have already appeared from his pen give much encouragement that he will succeed in working out this subject, and in giving us the underlying facts that have so long enshrouded the cause of caries of the teeth in mystery.

But in this we are anticipating. There is another work which demands examination before we pass to this. The views of Dr. Watt have received much commendation in this country, but not so in the old. There, other men have been prominent, and we find that while they have agreed in the main, there are important differences between them. As we have said, Dr. Watt has maintained that decay is caused by the acids, nitric, hydrochloric and sulphuric, with possibly others. In Europe the influence of these particular acids has been very generally denied; and the results attributed to other acids, as the lactic, acetic, and the group known as the organic acids. Among those that have examined this subject, none, perhaps, have attained a wider hearing than Magitot, of Paris. This gentleman published a work on this subject in 1868, in which he makes an extended examination of the subject, arriving at the conclusion that decay is caused by acids. These acids, however, are derived from the saliva through the process of fermentation. Dr. Magitot instituted a long series of experiments to determine the effects of the suspected acids on the teeth. This series of experiments show that most of the organic acids act very feebly on the teeth in the proportion of one to one thousand of water; and that in the proportion of one to one hundred they act quite energetically; so that the teeth submitted to their action will be completely decalcified within a few weeks or months. Most of this series of experiments were continued, however, for two years. The conclusion seems to be that caries may be produced by any of the group of acids that may be developed by the fermentation of the saliva; these are the lactic, acetic, butyric, etc.

Dr. Magitot states distinctly, that the agency of micro-

8

organisms in the production of these acids is admitted by him, but discusses this phase of the subject no farther. He makes no effort to determine to what extent these acids may be formed in the mouth. All of his experiments were tried out of the mouth, and no provision whatever was made to ascertain the effect that fermentation may have had on his solutions in the progress of his experimentation. This being the case, the only result of the experiments is the determination of the strength of the solutions of these different acids necessary to decalcify a tooth.

This, together with the facts obtained from other sources, showing that most of these acids are the products of certain fermentations that may go on in the mouth, gives much force to Dr. Magitot's conclusions.

Very soon after Dr. Magitot's work, in the same year, indeed, came the work of Lieber and Rottenstein, to which we have referred. The work seems to have been written for the express purpose of showing that decay of the teeth is caused by the life and development of the fungus known as leptothrix buccalis. In this the authors seem to have signally failed. They certainly make but little advance toward the demonstration of the parasitic theory of this affection. Indeed, they do not seem to have endeavored to show that this fungus does more than promote decay that has already become established.

They say, after having made an extensive examination of the life and growth of the leptothrix, " from what has been said, it results that two principal phenomena manifest themselves in the formation of dental caries, viz., the action of acids, and the rapid development of a parasitic plant, the leptothrix buccalis." They do not suppose the leptothrix buccalis capable in itself of attacking the teeth, if their condition be normal, but when their surfaces are once softened by acids, then the fungus may penetrate the portions thus softened and continue the destruction. Again, they say, "It seems that the fungi are not able to penetrate an enamel of normal consistency. The dentine itself, in its normal condition of density, offers great difficulties to their entrance, and we are not yet sure that the leptothrix could triumph over this resistance." Again, "We cannot decide at present if the leptothrix is able to penetrate sound dentine, when from any circumstance it happens to be denuded. . . . . . But, if the enamel or dentine are become less resistant at any point, through the action of acids, or if, at the surface of the dentine, a loss of substance has occurred, then the elements of the fungus can pass into the interior of the dental tissues, and produce by their extension, especially in the dentine, effects of softening and destruction much more rapid than the action of acids alone is able to accomplish. . . . . The participation of the fungus is constant in the progress of caries which has reached this stage. As soon as a loss of substance can be shown, there is found the presence of the fungus, so that the question whether or no the acids alone could produce ravages more considerable is without importance."

The *modus operandi* by which leptothrix may produce softening of the dentine is left without explanation. We can conceive, however, that they may do something to assist the softening process by the outpouring of a digestive fluid. If, however, this fungus gave a fluid that would digest a tooth, we would think that sound teeth would be very scarce, for it grows abundantly in every mouth.

Since the time of Lieber and Rottenstein's work we remember of no other of much importance having appeared on this subject. The discussion has continued, however, in the journals. We cannot now undertake to review this literature, interesting as it would be, but must content ourselves with one writer, Dr. Miller, now of Berlin.

Dr. Miller's experiments bear the stamp of being more carefully performed than any that have previously come to our knowledge. This was to have been expected, from the fact that they are the latest, giving the experimenter the advantage of all that has gone before; and for the reason that he is very favorably placed for such work, being in the midst of the best experimenters of the world. Therefore his work is looked to with unusual interest.

We need not, however, notice any but his last series of articles, that which is now appearing. We cannot, of course, criticise Dr. M.'s work now, for we have not heard him through ; but enough has appeared to show very clearly what the result will be.

Dr. M. begins this series of articles with this sentence: "During the last two years I have stated at different times and places, as the result of many experiments, that 'the first stages of dental caries consist in a decalcification of the tissues of the teeth by acids, which are, for the greater part, generated in the mouth by fermentation.' The object of the investigations described in this and the following papers, is to determine this ferment, and the conditions essential to its action."

We see from this, that Dr. Miller begins just where Dr. Magitot left off sixteen years ago. The discussion of the subject during these years has given us no additional facts, as to the essential nature of these phenomena; but the advance of thought in reference to the general subject of such investigations has been such that no man would now repeat Dr. Magitot's course of experimentation with the same end in view.

I will give a very brief synopsis of Dr. M.'s course. And while I do so, I wish you to keep in mind the object he has in view. It is admitted that decay is brought about by acids, developed, or Dr. M. supposes them to be developed by fermentation of some kind. The object of this course of experiments is to find and examine this supposed ferment.

It is not necessary that I should describe in detail all the apparatus with which the experimenter provided himself; it is sufficient to say that all the appliances for the prevention of error were used. The first question to determine was whether or not the ptyaline of the saliva could so change starch as to produce an acid. This question was soon decided in the negative. The starch was promptly changed into sugar, but here the reaction ceased; the fluid remained permanently sweet when the proper precautions were observed to prevent the ingress of germs. This proves that the acidifying power does not belong to the saliva. It must then be something foreign.

Now, a freshly extracted carious tooth was taken, all food removed, the outer portions of the decayed mass saturated with a ninety per cent. solution of carbolic acid, to destroy any accidental germs that might be in this portion. Then, with an instrument, purified by heat after each cut, layer after layer of the softened dentine was removed until the inner portions were reached. Then a slice was quickly conveyed to a sterilized culture medium, composed of sterilized saliva, water, sugar and starch, and placed in an incubator, together with another test-tube of the same culture fluid uninfected, to serve as a check. In twenty-four hours the infected culture became acid, while the other did not. This remained constant in a sufficient number of experiments to establish the fact that the acidity was due to the infection. From the cultures that had become acid, other cultures were infected, which also became acid, thus proving that the experimenter was dealing with a ferment that was capable of propagating itself; an organized ferment.

Microscopic examination showed that these cultures contained an organism similar to those found in the deeper layers of carious dentine, and which remained constant in their characters. Chemical examination, which seems to have been very carefully conducted, showed the acid produced to be lactic acid. This acid has been shown to be capable of decomposing the teeth by M. Magitot and many others. Yet Dr. Miller goes still further, and by placing sections of dentine in his culture fluids that he has infected, finds that they are decomposed by the acid formed; while such sections placed in such fluids not infected are not changed. Thus he not only proves that an acid is formed, but that the acid is formed in sufficient amount to destroy the dentine.

This, when compared with the best experimentation previously had, marks a great advance. One point seems to have been gained. One organism has been traced thus far, and may now be said to have been proven to be able to produce certain of the phenomena of decay. But this is not all: There is much yet to be done. True, one other point is spoken of by Dr. Miller. All who have made a careful study of caries know that there is a peculiar enlargement of the tubules, which is not seen in dentine softened by acids alone. Dr. Miller has been looking for this also, and not without success; for in some sections of dentine exposed to the action of the cultures, he found the organisms crowding into the tubules, and tells us that he also found them enlarged, as in natural caries of the teeth in the mouth; indeed, that he had before him veritable caries, artificially produced.

This delineation of results of experimentation must have great weight in the settling of the problems at issue; especially if they are confirmed by other competent observers. There is nothing in these experiments that is not in harmony with known facts, unless it be the widening of the tubules by the crowding in of the organisms. It is known, by previous experiment, that this widening is not caused by the lactic acid as it exists dissolved in the surrounding medium. And I think very few will be willing to concede that the organisms can accomplish this by physical force. This point requires further investigation, and its study will doubtless lead to further discoveries. However, we think it may be explained in advance; at least, the effort may serve to direct experimentation.

In a previous lecture we have dwelt at some length upon the digestive fluids of living organisms. Unexpectedly we find use for these ideas now, for they were written before we saw Dr. Miller's last article. I have explained how it is that dead bone, roots of the temporary teeth, ivory driven into the flesh, catgut ligatures, sponge, etc., are dissolved and removed by a soluble ferment. I have also shown that the soluble ferment of the yeast plant has been found and proven: also that of ammoniacal fermentation; and how plants take up otherwise insoluble substances. Now this widening of the tubules is a conceded fact. It is also shown that it is not done by the lactic acid in case of other experiments; nor can it be done by the physical force of the organisms; but it can, in all probability, be done by the digestive fluid of the organism. The conditions for this work of the digestive fluid is the same as that of the granulations in widening the meshes of the sponge and finally removing the last of it. As yet no such soluble ferment has been demonstrated in connection with this organism; but theoretically it must exist, and if Dr. Miller should undertake to search for it, he will be able to demonstrate it speedily, and determine its co-operation in producing some of the phenomena of decay. At least, determine its capabilities.* This organism cannot be said to

---

* Since the above was written Dr. Miller has also reported the finding of another micro-organism that he thinks capable of producing caries.

have been definitely and completely studied until this soluble ferment, or diastase, be found, isolated, and its capabilities separately determined.*

It is by no means probable that this is the only organism that may stand in a causative relation to caries. The organism of butyric fermentation, possibly that of acetic fermentation, and a large number of others of the acid fermentations may cause decay; nor is it by any means a settled fact that decay of the teeth may not be brought about in part by other vital processes than the acid fermentations. Of this, however, we will speak later.

Another question may arise in this matter, and need explanation. I have repeatedly said that the waste products of an organism prevented the activity of that organism, when collected in a certain amount. How, then, can this organism continue to thrive in its own waste product, and thus continuously promote caries by furnishing more, and still more, of this waste product? Simple enough. Every chemist who has studied lactic fermentation has been in the habit of introducing some form of lime into the fermenting fluid to "fix" the lactic acid in the form of a lactate of lime, in which case it does not hinder the progress of the fermentation. In this way a much larger amount of the lactic acid may be obtained, as it is readily regained from its salts. This was learned long before the organism was found. Now in the production of caries, the tooth presents the lime for the formation of the lactate, and thus furnishes the very conditions necessary to the continuous growth of the organism.

In this connection, I wish to call attention again to a difficulty that has ever existed in the study of the action of organized germs in producing impressions of whatever kind,

* Since the above was written Dr. Miller has reported the finding of this soluble ferment.

whether it be illness or the disorganization of structures. The power of the life force is always manifested in the remoleculization of matter. The chemical forces of matter are the playthings of the life force. The molecule is compelled to admit other elements, is split in twain, is torn to pieces for the formation of other molecules of a different character; and finally, it is cast aside, so changed in its molecular and physical characters that it is unrecognizable as the same, outside of the chemist's laboratory. Life is an immaterial force, but its dealings are with the material. In the study of the questions now before us, it is with these remoleculizations of matter that we have to deal. The mere physical power of these low organisms, while in some peculiar positions it may amount to something, may, as a rule, be left out of the count entirely.

If micro-organisms decompose an organic body, they do so by furnishing, through their remoleculizations, a chemical substance capable of acting on that body. And, furthermore, this substance may, and generally is, produced by the remoleculization of the elements of the organic body being decomposed. Bacterium lactis lives at the expense of starch or sugar.

If micro-organisms cause an illness, a fever, or an inflammation, it is because that organism, in its remoleculizations of matter, elaborates a toxic substance capable of producing these results. What is the source of the greater number of the toxic substances known and used in the medical world to-day? Take strychnia, opium, quinia, rhubarb, aconite, veratrum, atropia, nicotine, and a host of others; are they not the results of remoleculizations by the life force as manifested in plants? Then take alcohol, ammonia, succinic acid, propionic acid, lactic acid, and many other toxic and irritant substances, are they not the result of remoleculizations, brought

8*

about by the life force, as manifested in the form of micro-organisms? Does any one pretend that all of the toxic bodies elaborated by micro-organisms have been found, and their effects on the animal economy made out? Certainly not. Then we may expect to find others, if we make the search. It seems certain that every organism that is hurtful to man, is so by the elaboration of some substance that has toxic or irritant properties.

As we have said, we have no doubt that there are other micro-organisms than the bacterium lactis that may be instrumental in the production of dental caries. Nor do we think that it is necessarily only acid-producing organisms that may produce caries or some of its phenomena. This, however, need not be discussed now.

I wish now to turn your attention for a moment to another class of phenomena, and make some inquiry into their possible participation in the production of caries. I have already spoken of the strong probability that the otherwise normal tissues, when under the influence of certain qualities of inflammation, emit a fluid of acrid and very irritating properties. About the necks and other parts of children we sometimes see this fluid excoriating the skin wherever it touches it, seemingly acting the part of a caustic. I have already referred to the fact that dead bone, ivory driven into the flesh, sponge, catgut ligatures, etc., are dissolved and removed. This causes us to inquire whether or not a substance may be elaborated in the same manner in the mouth, that may have its quota of effect in the production of the phenomena of caries.

I was thoroughly convinced that this was the case years ago, though I endeavored to explain the results by the assumption that an acid was formed. This is entirely unnecessary. Soluble ferments do not seem to depend for their action on either acidity or alkalinity. They seem to be controlled by some other than the known chemical laws, and their action is not yet understood. We have no means of explaining them. If a piece of ivory thrust into the flesh is attacked and burrowed out in holes by a secretion thrown out by virtue of the irritation induced, as asserted by Krause, Kölliker and other of the most capable observers of the world, why may not a tooth be attacked in the same way, by virtue of an irritation of the tissues about its neck or during the irritation consequent upon its eruption when this is unusually prolonged? As a matter of fact, it has been observed that decays are very prone to occur in just such situations as tend to confirm this hypothesis. Thirteen years ago we drew attention to this in a paper before the Illinois State Dental Society. Both before and since that time I have given much attention to this point, and I am more than ever convinced that it has much to do with the beginnings of decay. I do not wish to be misunderstood in my view of this matter. It is not my notion that decays are initiated by this cause alone. It is only one of the first steps by which other forces which come later are rendered operative. The means, if you please, by which the surface of the tooth is first broken, and by which organisms are permitted to find a lodgment. Not a means by which a decay is carried on to the complete destruction of the tooth. This effect cannot be produced except while the tissues are in contact, or in very close proximity to the part in process of solution, for the reason that the secretion from the tissues would be dissipated in the fluids of the mouth before they could have time to produce their effects upon the tooth structure.

The positions at which these results are seen are—wisdom teeth, that come through very slowly; on the buccal surfaces of the molars generally; and sometimes on the labial surfaces of the upper incisors. In case of the wisdom teeth, the fact that they are very often decayed before they are fully through the gum is especially remarked ; and as a rule, if these decays are carefully noted at a very early period of their progress, it will be seen that they are different from other decays in several respects. It always has its beginning under the free margin of the gum. There is usually no change whatever in the appearance of the tooth; the eye discovers nothing. The surface seems normal, or, at most, the portion of the tooth appears rather whitish ; but on trial with the excavator the instrument will, apparently, break in through the enamel prisms, disclosing a cavity of very slight depth. It often happens that the enamel may be easily scraped away over a considerable space, as though it was so much chalk. The depth will present much variation; often it is only a part of the thickness of the enamel, at other times we may find it extending into the dentine, in which it forms a veritable cavity. If there is much depth, however, the characteristics will have assumed the more usual type.

Occasionally we see this character of decay (if it may be so called) in the grinding surfaces of the wisdom teeth; occasionally in the first molar also, where the tooth has come through very slowly, and the gum has been for a long time in a state of chronic irritation. It is characteristic of this effect that it is as often seen on the smooth surfaces of the teeth as in the pits and grooves. No imperfection is necessary to prepare the way for this manifestation.

As we have said, this takes place under the gum ; is covered by the gum. Now, as the tooth rises higher and the surface thus affected becomes exposed, these spots are prone to become the seat of true caries, with all of its usual manifestations. However, very often caries does not take place. In this case a whitish spot is seen, which gradually assumes a yellowish tinge, then brownish, and finally becomes black. This result is brought about by the settling into, or the formation in, the affected tissue of the black sulphurets; as I was the first to show. (See Report on Dental Chemistry, by Dr. H. A. Smith, Transactions of the American Dental Association, 1874, page 78.)

We see these spots every day upon the sides of the molars, in every stage of coloration, from ashy white to a deep black. They are as apt to be on the otherwise smooth surfaces as in the pits and grooves. They may occur on any of the teeth, but are oftenest seen in the positions named. The decays that so often occur on the labial surfaces of the upper incisors, are often, though not uniformly, of the same character.

Those decays that occur just at the junction of the enamel and cementum, in persons of middle age or past, also, occasionally, in younger persons, in very many cases, seem to be of the same character. Some irritation of the gum at the immediate spot seems to be one condition of their beginning. These also have some special characteristics not common to other decays. If they are closely examined very early, in their inception, it will be found that the cementum has been removed and that the margin of the enamel has become chalky. Soon after this, if the case continue to progress, the gum, which till now was closely applied to the part, becomes everted so as to expose the breach in the tooth. This often becomes the seat of the most exquisite sensitiveness just at the present stage of the process, which calls the attention of the patient to the spot. Generally, however, nothing can be seen by either patient or operator, except a slight eversion of the gum, and the slight grooved appearance of the neck of the tooth, which the operator is often puzzled to differentiate from the normal form of the tooth. However, if he will carefully press the gum away (and he will find this abnormally sensitive) until he can see the root of the tooth below plainly, it will be easy to demonstrate that there has been a decided loss of substance. Trial with an instrument will develop the fact that the surface within this groove is exceedingly sensitive; the dentine is exposed.

Now, if the case is left to itself, this sensitiveness will continue for some weeks, or even months, and then abate; and it will be found that the case has taken on the usual characters of decay. It may cease to progress and assume a dark color, or it may progress rapidly, and remain of a more or less ashy cast.

It has been our opinion that this class of decays, if that term can be applied at this stage, is brought about in precisely the same way that the root of a permanent tooth is partially absorbed on account of a chronic irritation of its peridental membrane. In other words, a soluble ferment has been called out by the irritation that has dissolved out a part of the tissue at that point. Or, if you prefer to have it put in that way, a true absorption has taken place which forms the nidus for the future decay.

Another class of decays are very common, which I have studied very closely, and which seem to be of the same character in their inception. These begin under the free margin of the gum, under plates that abut closely against the teeth. These are usually very rapid in their course, evidently for the reason that as soon as the free margin of the gum is everted, a pocket is formed by aid of the plate, in which fermentation can proceed to the very best advantage. It does not seem that the beginning of this decay is often after the eversion of the gum has uncovered the spot. Of course we often see decays occur where clasps encircle the teeth that are high up on the crown. Such must not be confounded with those that begin at the margin of the cementum.

A question of great interest here, and the findings of Dr. Miller make it doubly so, is the composition of the fluid that is instrumental in the absorption of bone. I believe that I have already referred to the opinion of Krause, and others who regard it as containing lactic acid as its active principle. Krause bases this opinion on the behavior of the tissue forming this substance toward staining agents. From his experiments, it seems quite certain that the substance has an acid reaction, but the particular acid seems to me to be undetermined. The action that we see in the absorption of the roots of the teeth is not the action of lactic acid alone. This acid dissolves the lime salts only, leaving the tissues of the tooth behind in the normal form, as has been well shown by Magitot and many other capable experimenters.

This, of course, excludes the lactic acid as the exclusive agent in the work. Nevertheless, the secretion may contain this acid in combination with some other active principle.

INDEX.

www.ingramcontent.com/pod-product-compliance
Lightning Source LLC
Chambersburg PA
CBHW022103210326
41518CB00039B/457